Irradiation of Dry Food Ingredients

Author

József Farkas, D.Sc.
Director
Institute of Food Technology
University of Horticulture and Food Industry
Budapest, Hungary

CRC Press, Inc.
Boca Raton, Florida

Library of Congress Cataloging-in-Publication Data

Farkas, Jozsef, Dr.
Irradiation of dry food ingredients

Includes bibliographies and index.
1. Food, Dried. 2. Radiation preservation of food.
I. Title.
TP371.5.F37 1988 664'.0284 87-20871
ISBN 0-8493-6686-0

Direct all inquiries to CRC Press, Inc., 2000 Corporate Blvd., N.W., Boca Raton, Florida, 33431.

International Standard Book Number 0-8493-6686-0

Library of Congress Card Number 87-20871
Printed in the United States

FOREWORD

During the last 15 years, increasing concern over the quality of environment, consumer safety, and occupational hazards has resulted in major direct impacts on industrial processes. This can be observed in the area of disinfestation and microbial decontamination of dry commodities of agriculture and food industry. Commodities requiring decontamination in the dry state can be classified as raw materials (flours, etc.) or minor ingredients (spices, etc.) used by the food industry, or foods for direct-consumption (dehydrated soups, etc.). There is a growing opposition to undesirable chemical residues and interaction products of chemical treatments of food. Research and development over the past 30 years have proved on a large variety of foods and feeds that treatment with ionizing radiation (γ-rays, accelerated electrons, or X-rays) is a viable alternative for destroying contaminating organisms.

Ionizing radiations do not cause any practical rise in temperature; they produce only negligible, minute amounts of radiolytic products, and at the same time they are very efficient microbicidal and insecticidal agents. Radiation treatment can be applied to hermetically packaged products, thereby excluding recontamination. Radiation processing is a direct, simple, and highly energy-efficient treatment.

Food irradiation is not a panacea, it is a case-by-case matter, but it holds promise for improving both the quality and shelf-life of many food products with efficacy and safety. As a part of the worldwide food irradiation research and development efforts, decontamination of dry ingredients by ionizing radiation has developed into one of the food irradiation processes with the most immediate application potential. Advantages of radiation processing include reliability and simplicity of process control, and a high standard of safety.

This review includes a wealth of data obtained on irradiation of dry ingredients, herbs, and enzyme preparations in various countries for reducing microbial contamination and preventing insect damage in such commodities. Proper regulatory action, and political and socioeconomic feasibility can produce a major radiation application area, following the well-established applications of radiation treatment to sterilizing packaging materials, disposable medical products for hospital and home health care, disinfesting sewage sludge, crosslinking of heat-shrinkable plastic film, tubing and insulation of electrical wire and cable, and curing polymeric coatings and rubber goods.

THE AUTHOR

József Farkas, D.Sc., is professor of Food Technology and Director of the Institute of Food Technology, University of Horticulture and Food Industry, Budapest, Hungary.

Professor Farkas obtained his B.Sc. (Dipl.Ing.) degree in 1956, the M.Sc. (Dr.tech.) in food technology in 1964 from the University of Polytechnics, Budapest. He received his Ph.D. and D.Sc. in radiation microbiology in 1968 and 1979, respectively, from the Hungarian Academy of Sciences, Budapest. He joined the Central Food Research Institute, Budapest, in 1957 and served as Scientific Officer, Head of the Microbiology Department and Food Irradiation Section, then as Deputy Director until 1986, when appointed to the University of Horticulture and Food Industry. He worked as Research Scholar at the Federal Research Institute for Food Preservation, Karlsruhe, West Germany, in 1959, at the ARC Meat Research Institute, Langford, Bristol, U.K. in 1968 and at the Biophysics Laboratory, Department of Biology, Illinois Insititute of Technology, Chicago, in 1972. He served in numerous FAO/IAEA panels and in 1973 as IAEA technical assistance expert at the Department of Agriculture and Biology, Nuclear Research Centre, Tuwaitha, Iraq. He was also Director of the International Facility for Food Irradiation Technology, an international project jointly sponsored by the FAO, the IAEA, and the Dutch government, where he organized five interregional training courses on food irradiation from 1980 to 1985.

Professor Farkas has taught several graduate and postgraduate courses in food microbiology and food technology. He is secretary of the Joint Complex Committee on Food Science of the Hungarian Academy of Sciences and the Ministry of Agriculture and Food. He is a board member of the Hungarian Society of Microbiology and the Section of Microbiology, Biotechnology and Hygiene of the Hungarian Scientific Society for Food Industry, and member of the editorial board of *Acta Alimentaria*. He is chairman of the Food Irradiation Working Group of the European Society of Nuclear Methods in Agriculture (ESNA).

His research activities have covered a broad field of research and development from radiation biology of food-borne microorganisms and effects of physical and chemical agents on bacterial spores to the technology and economics of food irradiation. His team established the first Food Irradiation Pilot Plant in Hungary in 1970. He is author or co-author of seven books and has authored over 160 research papers and 110 scientific journal articles. His accomplishments were recognized by the Sigmond Award of the Hungarian Society for Food Industry.

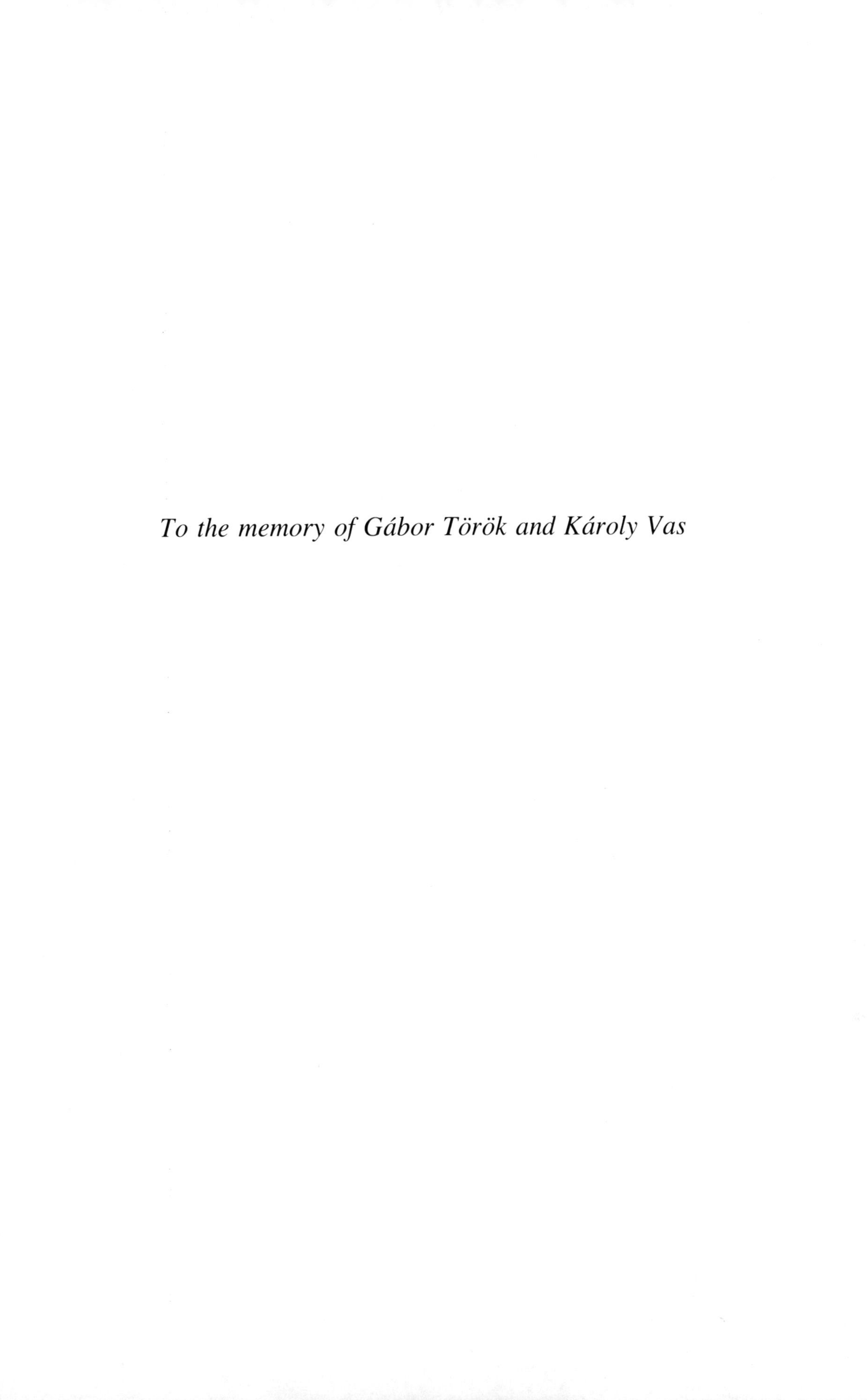

To the memory of Gábor Török and Károly Vas

TABLE OF CONTENTS

Chapter 1

INTRODUCTION

I. MICROBIAL CONTAMINATION OF DRY INGREDIENTS AND ITS SIGNIFICANCE TO THE FOOD INDUSTRY AND PUBLIC HEALTH

The rate of microbiological deterioration of a composite food usually depends on the degree of contamination of its constituents. In many cases, an important source of microbial contamination may be one of the minor components of the product. Under the prevailing production and handling practices, many food ingredients, herbs, and processing aids, i.e., spices, condiments, texturizing agents, enzyme preparations, etc., contain a large number of microorganisms which may cause spoilage, defectiveness in foods, or, more rarely, disease. The bacteriological examination of 357 specimens of 47 different untreated dried food products in Germany in the period 1980 to 1983 revealed a *Salmonella* contamination of 29 samples (8.1%).[66]

Microbial counts in dry ingredients can vary from one storage period to another depending on moisture content and storage conditions. The observed count is a function of original bioload, proliferation, and die-off. Counts frequently decrease during storage.

The spoilage and/or health hazards presented by an ingredient must always be evaluated in the context of its use. Use, for example, of spices that may contain heat sensitive pathogenic bacteria may be of no consequence if used in a food that is to be canned. The same spices — when used in products not to be cooked, and which would support growth, and are subjected to abusive temperature — clearly introduce a health hazard. Elevated aerobic, mesophilic spore populations in ingredients to be used in low acid products that undergo a mild heat treatment, i.e., canned pasteurized luncheon meats or canned cured hams, may cause spoilage if mishandling occurs in marketing channels or by consumers. For the canning industry the contamination of ingredients with heat resistant bacterial spores is especially troublesome. In some canned foods, the microbial flora of the major components consists of relatively heat sensitive organisms. If, however, a certain minor ingredient contributes a completely heat resistant microflora in terms of the process being applied, then this item becomes the one important determinant of the bacteriological quality of the product, and control over the finished product involves control over the spore load of this ingredient.[1] Spices and cereals are, for instance, potentially prolific sources of heat resistant spores of bacteria, including thermophilic flat sours, putrefactive anaerobes, and sulfide stinkers.[2,3]

The destruction of thermoduric bacterial spores introduced to the food by ingredients often requires a severe heat treatment, which ensures the microbiological stability only at the cost of substantial reduction of nutritional and sensory quality of the manufactured product. Thus, a major concern of food processors is that the microbial load of ingredients and processing aids does not contribute to spoilage of food and does not diminish its microbial safety. Even when there is a low probability of contaminated ingredients causing spoilage (e.g., in dry gravy bases or dehydrated soups) ingredients may introduce numbers of microorganisms considered undesirable to industrial and regulatory interests. Dry soups and other dry convenience foods contain a number of dehydrated ingredients (cereals, protein concentrates, dried eggs, sugar, flavorings, dried dairy products, etc.). The microbiology of the mix becomes that of the ingredients.[4]

Various microbiological limits for foods and food additives are presented as suggestions and recommendations by noncommercial organizations, as specifications in sales contracts by buyers or as regulations imposed by the government. Microbiological limits for some dry food ingredients recommended by the International Commission on Microbiological

Table 1
MICROBIOLOGICAL LIMITS FOR SOME DRY FOOD INGREDIENTS RECOMMENDED BY THE ICMSF[68]

Product	Test	Limit/g[a]			
		n	c	m	M
I. Animal origin					
Egg products[b]	Standard plate count (SPC)	5	2	10^4	10^6
	Salmonella[c]	10	0	0	—
Enzymes	*Salmonella*	10	0	0	—
Meats or components including gelatin and fish protein concentrate	*Clostridium perfringens*	5	1	10^2	10^4
	Staphylococcus aureus	5	1	10^2	10^4
	Salmonella[c]	10	0	0	—
II. Cereal origin					
Cereal byproducts (bran, flours, etc.)	Molds	5	2	10^2	10^4
	Spores of rope-forming bacteria	5	2	10^2	10^4
	Bacillus cereus	5	1	10^3	10^5
	C. perfringens	5	1	10^2	10^4
III. Fruit-based					
Fruits, sun-dried	Osmophilic yeasts	5	2	10	10^3
	Molds	5	2	10^2	10^4
	Escherichia coli	5	2	<3[d]	10
IV. Vegetable origin					
Cocoa	SPC	5	2	10^4	10^6
Coconut	Molds	5	2	10^2	10^4
	Coliforms or Enterobacteriaceae	5	2	10	10^3
	Salmonella[c]	60	0	0	—
Dyes	SPC	5	2	10^4	10^6
Enzymes	*E. coli*	5	2	<3[d]	10
Gums	SPC	5	2	10^4	10^6
	Coliforms or Enterobacteriaceae	5	2	10	10^3
Nuts	Molds	5	2	10^2	10^4
	E. coli	5	2	<3[d]	10
Spices	SPC	5	2	10^4	10^6
	Molds	5	2	10^2	10^4
	E. coli	5	2	10	10^3
Vegetables	*E. coli*	5	2	<3[d]	10^2
	Salmonella[c]	10	0	0	—

[a] From n samples analyzed c samples may exceed m, but none may exceed M.
[b] Do not apply to egg albumin desugared by bacterial fermentation.
[c] All *Salmonella* tests are subject to compositing of the sample units.
[d] 3 means no positive tube in the standard 3-tube MPN technique.

Specifications for Foods (ICMSF) are listed in Table 1, and advisory microbiological end-product specifications for dried milk, edible rennet casein, and food grade whey powders published by the International Dairy Federation are given in Table 2.[68,70]

Concerning microbiological specifications, it was rightly pointed out by Christensen[69] that "while it is often possible and desirable to establish a minimum standard for microbiological *safety* of food or ingredients, it is less obviously useful to establish a minimum *quality* standard. This is because a number of quality grades may be used in the fabrication of different foods and there will be a nexus between quality and cost."

Table 2
ADVISORY MICROBIOLOGICAL END-PRODUCT SPECIFICATIONS FOR DRIED MILK, EDIBLE RENNET CASEIN, AND FOOD GRADE WHEY POWDER PUBLISHED BY THE INTERNATIONAL DAIRY FEDERATION[70]

	Limits/g			
Microorganisms	**n**	**c**	**m**	**M**
Salmonellae[a]	15 (× 25 g)	0	0	—
Mesophilic aerobic bacteria	5	2	5×10^4	2×10^5
Coliforms	5	1	10	10^2
Staphylococcus aureus	5	1	10	10^2

[a] Not detectable in 15 samples of 25 g each.

The practical control of microbial contamination of food products by organisms from ingredients would lie logically in the selection of materials relatively free from significant microorganisms. The work involved in carrying out a routine bacteriological examination of all ingredients is, however, hardly within the capacity of all but a few of the largest laboratories. Some ingredients may inherently carry troublesome microorganisms, in spite of all efforts to improve their production conditions. All these problems account for attempts to reduce the viable cell counts of these ingredients by decontamination treatment. In most of the cases, it is unnecessary to achieve full sterility; only a proper reduction of the viable cell count is needed, as the ingredient is incorporated into unsterilized food.

The appropriate decontamination treatment

1. Should be carried out safely and fast
2. Should be effective against all microorganisms
3. Should be able to penetrate the packaging and product in order to act against all the microorganisms present
4. Should be adaptable to large quantities of material with high efficiency
5. Must not reduce the sensory and technological qualities of the treated commodities

All of this must be accomplished without affecting the health of the consumer of the decontaminated product and at treatment costs which are acceptable to the user.

II. CHEMICAL METHODS OF DECONTAMINATION AND THEIR LIMITATIONS

Until recently, fumigation with volatile microbicide compounds, particularly with ethylene oxide (EtO), has been widely used for decontamination and disinfestation of ingredients and various dried foods.[5,7] According to the American Spice Trade Association, in the U.S. alone, 800,000 lb of EtO were used in 1977 for treatment of 80,000,000 lb of spice.[8] Actually, because of the extreme flammability of EtO, various nonflammable mixtures of EtO oxide in inert gases (carbon dioxide or a chlorinated hydrocarbon) are applied in vacuum. These inert gases do not add to or detract from the biocidal activity of EtO. The concentration range of EtO used in treatments to control microorganisms in dry food commodities is 400 to 1000 mg/ℓ.[9] The fumigation treatments are normally carried out in specially designed vacuum chambers.

Another epoxide, propylene oxide (PO) has been suggested, and, on a more limited scale than EtO, used for "gaseous sterilization". Some countries, however, did not allow the use of PO for foods, and this compound was found to be much less active and less penetrating than EtO in powdered and flaked foods.[10] Concentrations of 800 to 2000 mg/ℓ are necessary for PO treatments.[9]

Methyl bromide, a widely used insecticidal fumigant, has little impact on the microbial population of dried products when it is used at accepted levels.

Even though EtO is one of the most efficient gaseous antimicrobial agents, fumigation is a time-consuming batch procedure. At a temperature of 20° to 25°C, fumigation for 6 to 7 hr is required, though the requirement varies according to the microflora present.[11,12] Fumigation appears to be beset with problems of uniformity of decontamination and unpredictable levels of reduction in the microbial counts and requires a complex process monitoring.[13,14] The rate of destruction of microbial cells depends on the concentration of the fumigant, temperature, relative humidity of the atmosphere in the fumigation chamber, degree of dryness of the microbial cells, amount of organic matter present, the porosity of the product, and the permeability of the packaging material.[15,16] In several cases fumigation is not efficient enough with commodities of very low moisture content. In the case of onion powder, treatment with EtO is a costly and labor intensive procedure since the dried, powdered onions must be dried and reground subsequent to treatment, as the onion powder clumps at the humidity level required for fumigation. Slight changes in color and flavor of some fumigated ingredients were noted as well.[17,19] Fumigation of mustard and mustard-flour is hindered by the strong off-odor produced.[20] A significant decrease in the volatile and nonvolatile oil content of black pepper and allspice by EtO was reported.[21,22]

More serious problems pertinent to fumigation are the health-related ones. The fumigants may be physically retained for some time by the fumigated commodities, and both EtO and PO, being strong alkylating agents, may undergo chemical reactions with many food components.[11,23] A few of the typical reactions of EtO are

Reactant	Product
Water	Ethylene glycol
Halides	Ethylene halohydrins
Alcohols and phenols	Ethylene glycol ethers
Acids	Ethylene glycol esters
Amines	Ethanolamines
Sulfhydryl compounds	Thioethers

The duration of exposure to the fumigant, its concentration, the permeability of materials to EtO, and their moisture content and temperature are some of the main factors which determine the extent to which these reactions can go.

The absorptively bound EtO residues may be fairly high after gassing, even after repeated evacuation.[24] Although these EtO residues decrease continuously during storage of the fumigated product, this is frequently due to further chemical reactions in the fumigated commodity rather than to the loss of the gas. This is clearly shown in Figure 1.[25]

Wesley et al.[26] were the first to demonstrate that in the presence of water and inorganic halides, ethylene chlorohydrin (ECH) and ethylene bromohydrin (EBH) may be formed (in addition to ethylene glycol) in fumigated products. EBH is formed by the reaction of EtO with inorganic bromide, introduced by previous treatment of the commodity with methyl bromide.[27] Although the persistence of the ethylene halohydrin residues during storage varies considerably from one substrate to another, in general, they are much more persistent than EtO and may also persist under food processing conditions. No decrease in ECH residue was observed within 4 months in black pepper and whole turmeric.[28] de Boer and Janssen[29]

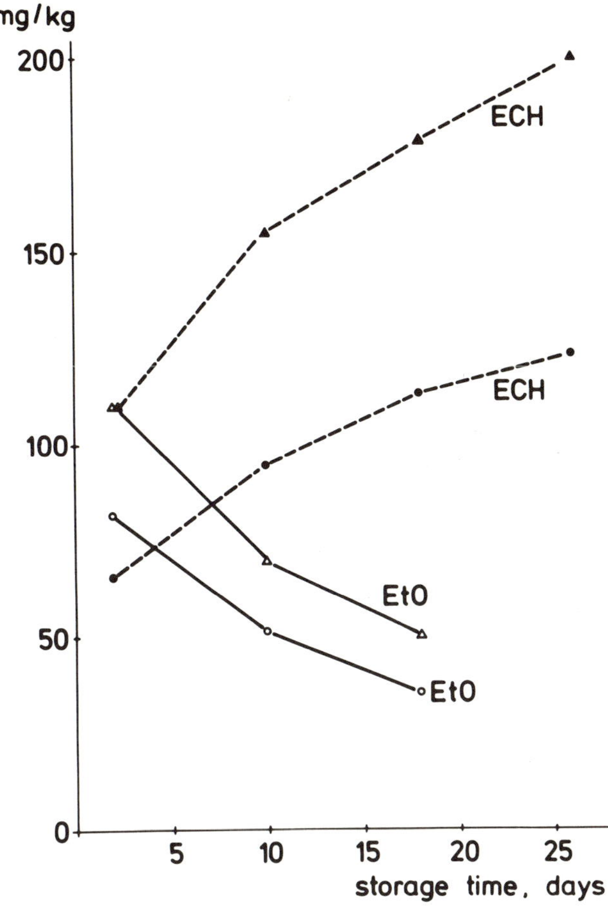

FIGURE 1. Change in EtO and ECH content of fumigated black pepper during storage. Fumigation with 500 g EtO per cubic meter for 6 hr (○ and ●) and 18 hr (△ and ▲), respectively.[25]

measured on the average about 40% reduction of ECH residues (from 525 to 313 ppm) after 2 months of storage in 27 fumigated spice samples, and at the same time, the average concentration of ethylene glycol increased by about the same proportion (from 153 to 212 ppm). No wonder then that ECH was frequently found in EtO-treated ingredients, particularly in spices, in concentrations exceeding even the 1000 mg/kg level (see Table 3), although in several countries the maximum permissible level of ECH is 300 mg/kg. The LD_{50} dose of ECH for animals by oral administration is given as 7.2 mg/100 g.[32]

EtO also may form toxic 2-hydroxyethyl compounds with lysine and cysteine. Up to 30 ppm of these compounds was found in dry egg and milk powder.[8]

Both EtO and PO as well as the halohydrins are mutagens, and they are suspected of causing other chronic or delayed toxic effects.[8,33-43]

The effectiveness of EtO as a fumigant lies in its inherent toxicity; consequently, there is an acute concern regarding its toxic effects on exposed workers. At concentrations below 50 ppm as an 8-hr time-weighted average, which is the current exposure limit in several countries, EtO is suspected of being a carcinogen.[43]

Due to the above-mentioned problems, fumigation represents a significant occupational health hazard for workers in fumigation plants.[43-46] The presence of residues or reaction products gives rise to more and more toxicological misgivings. The mounting concern in

Table 3
LEVELS OF ECH IN COMMERCIALLY AVAILABLE SPICES[a]

Country and sampling	Percentage incidence of samples with ECH levels (mg/kg)		Ref.
	>300[b]	>1000	
Switzerland	60	19	28
From 58 samples of 16 spices			
Sweden	49	24	30
From 100 samples of 28 spices			
Netherlands	35	6	29
From 34 samples of various spices			

[a] Summarized from References 28 to 30.
[b] Proposed permissible level for ECH residues in spices is 300 mg/kg according to the British Industrial Biological Research Association.[31]

recent years has already resulted in a number of restrictions in the permitted uses of fumigants, and more and more regulatory restrictions are coming into power.[8,47,48]

On the basis of recent mutagenicity and carcinogenicity studies, the U.S. Occupational Safety and Health Administration in 1984 reduced the air exposure standard for EtO from 50 to 1 ppm determined as an 8-hr time-weighted average concentration.[43] An ''action level'' of 0.5 ppm is established as the level above which employers must initiate periodic employee exposure monitoring and medical surveillance. The National Institute of Occupational Safety and Health (NIOSH) has recommended lowering this standard further to 0.1 ppm.[49]

The use of EtO for all foods has been banned, for example, in Switzerland and in Japan as of 1982.[50,51] In the Federal Republic of Germany, fumigation with EtO as an insecticide treatment was forbidden in 1980.[53] Since October 1983, the responsible health authorities of the country no longer consider EtO as a compound which ''disappears'' from the treated product (Verschwindestoff).[54]

As an alternative ''gaseous sterilization'', treatment of natural spices and other foods with hot ethanol vapor was recently suggested.[55] The limitations of this treatment are, however, obvious. The systematic studies in the Federal Meat Research Institute, Kulmbach, West Germany have shown that the hot alcohol vapor treatment may result in the desired microbiological effect with whole seeds, i.e., whole pepper, but it is not feasible for ground or leafy spices.[56]

Another chemical method is the reduction of viable cell counts by acidification or *in situ* salt formation.[57-59] This treatment is applicable only over a very limited range of conditions and commodities.

Treatment with hydrogen chloride gas or application of glutaraldehyde and propionaldehyde, respectively, as well as various solvents (ethanol, methanol, acetone, or dichloromethane) were unable to significantly reduce the viable cell count of whole spices, and most of them caused considerable adverse changes in sensory properties.[48]

Summing up this review, it seems highly desirable to introduce an appropriate physical method for decontamination.

III. PHYSICAL METHODS OF DECONTAMINATION WITH LIMITED UTILITY

Mechanical cleaning or sonication did not result in a significant reduction of the viable cell count in whole species and herbs.[48,60]

Because of heat sensitivity of the delicate flavor and other essential components of species and herbs or the specific functional properties of enzyme preparations or texturizing agents, a regular heat sterilization cannot be applied.[61,62] Microwave treatment showed no utility either, because microwave heating is seriously hindered at the low moisture content of dry commodities.[21,22,48] Due to the inhomogeneities of products and the dielectric field, the heating effect is very uneven in these materials, resulting in significant adverse changes in sensory quality.[48] Recently, a process utilizing superheated steam created some interest in the treatment of spice seed, berries, and roots or rhizomes.[67]

Another thermal method known to decontaminate spices is to treat them in an extruder.[67] Different pressure/time/temperature combinations have been tested; the method, however, does not seem to be in commercial operation.

UV radiation has also been tried; however, it is considered to be unsuited to bringing about effective decontamination of dry ingredients since its penetrating capacity is too low.[63-65]

REFERENCES

1. **Silliker, J. H.,** Total counts as indexes of food quality, in *Microbiological Quality of Foods,* Slanetz, L. W., Chichester, C. O., Gaufin, A. R., and Ordal, Z. J., Eds., Academic Press, New York, 1963, 102.
2. **Krishnaswamy, M. A., Patel, J. D., and Parthasarathy, N.,** Enumeration of microorganisms in spices and spice mixtures, *J. Food Sci. Technol.,* 8, 191, 1971.
3. **Saint-Lebe, L. and Berger, G.,** Preservation of powdered food products by irradiation: advantages and possibilities of the method: application to the case of maize starch, in *Proc. 4th U.N. Int. Conf. on Peaceful Uses of Atomic Energy,* Vol. 12, United Nations, New York, 1972.
4. **Karlson, K. E. and Gunderson, M. F.,** Microbiology of dehydrated soups. II. "Adding machine" approach, *Food Technol.,* 19(1), 86, 1965.
5. **Hall, L. A.,** Sterilizaton of spices by treatment with ethylene oxide, *Food Packer,* 32(12), 26, 1961.
6. **Szabó, P.,** Kisérletek a füszerpaprika csiraszámának csökkentésére (Experiments for reduction of microbial cell count in paprika) (in Hungarian), *Konzerv. Paprikai.,* 128, 1968.
7. **Mayr, G. E. and Suhr, H.,** Preservation and sterilization of pure and mixed spices, in *Proc. Conf. Spices,* Tropical Products Institute, London, 1972, 201.
8. **Gerhardt, U. and Ladd Effio, J. C.,** Äthyleneoxidanwendung in der Lebensmittelindustrie. Ein Situationsbericht über "Für und Wider", *Fleischwirtschaft,* 62, 1129, 1982.
9. The International Commission on Microbiological Specifications for Foods, Gases as preservatives, in *Microbial Ecology of Foods,* Vol. 1, Silliker, J. H., Elliot, R. P., Baird-Parker, A. C., Bryan, F. L., Christian, J. H. B., Clark, D. S., Olson, J. C., Jr., and Roberts, T. A., Eds., Academic Press, New York, 1980, 170.
10. **Bruch, C. W. and Koesterer, M. G.,** The microbicidal activity of gaseous propylene oxide and its application to powdered or flaked foods, *J. Food Sci.,* 26, 428, 1961.
11. **Coretti, K. and Inal, T.,** Rückstandprobleme bei der Kaltentkeimung von Gewürzen mit T-Gas (Äthylenoxyd), *Fleischwirtschaft,* 5, 599, 1969.
12. **Hadlok, R. and Toure, B.,** Mykologische und bakteriologische Untersuchungen entkeimter Gewürze, *Arch. Lebensmittelhyg.,* 24(1), 20, 1973.
13. **Frohnsdorff, R. S. M.,** Sterilization of medical products in Europe, *Rev. IRE,* 5(2), 7, 1981.
14. **Tsuji, K.,** Low-dose cobalt-60 irradiation for reduction of microbial contamination in raw materials for animal health products, *Food Technol.,* (2), 48, 1983.
15. **Hoffman, R. K.,** Toxic gases, in *Inhibition and Destruction of the Microbial Cell,* Hugo, W. B., Ed., Academic Press, London, 1971, 225.
16. **Russel, A. D.,** The destruction of bacterial spores, in *Inhibition and Destruction of the Microbial Cell,* Hugo, W. B., Ed., Academic Press, London, 1971, 575.

17. **Coretti, K.,** Kaltentkeimung von Gewürzen mit Äthyolenoxyd, *Fleischwirtschaft,* 9, 183, 1957.
18. **Lerche, L.,** Vorkommen und Undschädlichmachung von Salmonellenbakterien in Trockeneiweiss, *Arch. Lebensmittelhyg.,* 8, 267, 1957.
19. **Glas, A.,** *Praktisches Handbuch der Lebensmittel,* Bayerischer Landwirtschaftsverlag, Munich, 1965.
20. **Coretti, K.,** Sterilisierung von Gewürzen, *Fleischwirtschaft,* 58, 1239, 1978.
21. **Vajdi, M.,** Comparative Effects of Ethylene Oxide, Gamma Irradiation and Microwave Treatments on the Control of Microorganisms in Selected Spices, M.Sc. thesis, University of Manitoba, Winnipeg, 1970.
22. **Vajdi, M. and Pereira, N. N.,** Comparative effects of ethylene oxide, gamma irradiation and microwave treatment on selected spices, *J. Food Sci.,* 38, 893, 1973.
23. **Scudamore, K. A. and Heuser, S. G.,** Ethylene oxide and its persistent reaction products in wheat flour and other commodities. Residues from fumigation or sterilization and effect of processing, *Pest. Sci.,* 2(2), 80, 1971.
24. **Kröller, E.,** Untersuchungen zur Begasung von Lebensmitteln mit Äthyleneoxyd und zu dessen Restmengen-Bestimmung, *Dtsch. Lebensm. Rundsch.,* 102, 227, 1966.
25. **Gerhardt, U. and Ladd Effio, J. C.,** Rückstandsverhalten von Äthylenoxid in Gewürzen, *Fleischwirtschaft* 63, 606, 1983.
26. **Wesley, F., Rourke, B., and Darbishire, O.,** The formation of persistent toxic chlorohydrins in foodstuffs by fumigation with ethylene oxide and with propylene oxide, *J. Food Sci.,* 30, 1037, 1965.
27. **Heuser, S. G. and Scudamore, K. A.,** Formation of ethylene bromohydrin in flour and wheat during treatment with ethylene oxide, *Chem. Ind.,* 1054, 1969.
28. **Stijve, T., Kalsbach, R., and Eyring, G.,** Determination and occurrence of ethylene chlorohydrin residues in foodstuffs fumigated with ethylene oxide, *Trav. Chim. Aliment. Hyg.,* 67, 403, 1971.
29. **deBoer, E. and Janssen, F. W.,** Microbiology of spices and herbs (in Dutch), presented at Dutch Symp. on Food Microbiol., Delft, November 17, 1983.
30. **Gustafsson, K. H.,** Rester av ethylenklorhydrin i vissa importeradle industrikryddor (Residues of ethylene chlorohydrin in foodstuffs and spices intended for industrial manufacturing) (in Swedish), *Var Foda,* 33(1), 15, 1981.
31. British Industrial Biological Research Association, BIBRA announcement, *BIBRA Inf. Bull.,* 5, 644, 1966.
32. **Goldblatt, M. W. and Chiesman, W. E.,** Toxic effect of ethylene chlorohydrin. I. Clinical, *Br. J. Ind. Med.,* 1, 207, 1944.
33. **Brem, H., Stein, A. B., and Rosenkranz, H. S.,** The mutagenicity and DNA-modifying effect of haloalkanes, *Cancer Res.,* 34, 2576, 1974.
34. **Rosenkranz, S., Carr, H. S., and Rosenkranz, H. S.,** 2-haloethanols. Mutagenicity and reactivity with DNA, *Mutat. Res.,* 26, 367, 1974.
35. **Kucerova, M., Zhurkov, V. S., Polivkova, Z., and Ivanova, J. E.,** Mutagenic effect of epichlorohydrin. II, *Mutat. Res.,* 48, 355, 1977.
36. WHO-IARC, *Cadmium, Nickel, Some Epoxides, Miscellaneous Industrial Chemicals and General Considerations on Volatile Anesthetics,* Vol. 11, WHO-IARC Monographs on Evaluation of Carcinogenic Risk of Chemicals to Man, International Agency for Research on Cancer, Lyon, 1976.
37. **Embree, J. W., Lyon, J. P., and Hine, C. H.,** The mutagenic potential of ethylene oxide using the dominant lethal assay in rats, *Toxicol. Appl. Pharmacol.,* 40, 261, 1977.
38. **Hogstedt, C., Malmquist, C. N., and Vadman, D.,** Leukemia in workers exposed to ethylene oxide, *JAMA,* 241, 1132, 1979.
39. **Pfeiffer, E. H. and Dunkelberg, H.,** Mutagenicity of ethylene oxide and propylene oxide and of the glycols and halohydrins formed from them during the fumigation of foodstuffs, *Food Cosmet. Toxicol.,* 18, 115, 1981.
40. **Ehrenberg, L. and Hussain, S.,** Genetic toxicity of some important epoxides, *Mut. Res.,* 86, 1, 1981.
41. **Dunkelberg, H.,** Kanzerogene Aktivität von Äthylenoxid und seinen Reaktionsprodukten 2-Chlorethanol, 2-Bromethanol, Ethylenglycol und Diethylenglycol. I. Kanzerogenität von Ethylenoxid im Vergleich zu 1,2-Propylenoxid bei subacutaner Applikation an Mäusen, *Zentralbl. Bakteriol. Parasitenkd. Infektionskr. Hyg. Abt. 1 Orig. Reihe B,* 174, 383, 1981.
42. **Kligerman, A. D., Erexson, G. L., Phelp, M. E., and Wilmer, J. L.,** Sister-chromatid exchange induction in peripheral blood lymphocytes of rats exposed to ethylene oxide by inhalation, *Mutat. Res.,* 120, 37, 1983.
43. Occupational Safety and Health Administration, Occupational exposure to ethylene oxide, *Fed. Regist.,* 49(122), 25734, 1984.
44. **Calleman, C. J., Ehrenberg, L., Janssen, B., Osterman-Golkar, S., Segerbäck, D., Swenson, K., and Wachmeister, C. A.,** Monitoring and risk assessment by means of alkyl groups in hemoglobin in persons occupationally exposed to ethylene oxide, *J. Environ. Pathol. Toxicol.,* 2, 427, 1978.
45. **Osterman-Golkar, S.,** Tissue doses in man: implications in risk assessment, in *Developments in the Science and Practice of Toxicology,* Hays, A. W., Schnell, R. C., and Miya, T. S., Eds., Elsevier, Amsterdam, 1983.

46. **Anon.**, Ethylene oxide limits ordered by US judge, *Chem. Mark. Repr.*, January 10, 1983.
47. Kommission der Europäischen Gemeinschaften, Generaldirektion Landwirtschaft, Wissenschaftlicher Ausschuss für Schädlingsbekämpfungsmittel, Bericht des Ausschusses über die Verwendung von Äthylenoxid als Begasungsmittel für Lebensmittel 3532/VII/79-D endg. (Stellungnahme vom 19. Dez. 1979).
48. **Neumayr, L., Promeuschel, L., Arnold, I., and Leistner, L.**, *Gewürzentkeimung, Verfahren und Notwendigkeit,* Abschlussbericht für die Adalbert-Raps-Stiftung zum Forschungsvorhaben, Institut für Fleischforschung, Kulmbach, 1983.
49. **Anon.**, NIOSH recommends lowering air exposure standard for ethylene oxide, *Food Chem. News,* August 1, 1983, 56.
50. **Grünewald, Th.**, *Untersuchungen zur Bestrahlung von Trockenprodukten,* BFE-R-84-02, Berichte der Bundesforschungsanstalt für Ernährung, Karlsruhe, July 1984.
51. **Kawabata, T.**, FAO/IAEA Research Coordination Meeting on Asian Regional Cooperative Project on Food Irradiation (RPFI) and Third Meeting of the RPFI Project Committee, Bangkok, November 22 to 26, 1982, Ref. 52.
52. **Byrn, M. W., Kwon, J. H., and Cho, H. O.**, Sterilization and storage of spices by irradiation. I. Sterilization of powdered hot pepper paste (in Korean), *Korean J. Food Sci. Technol.*, 15, 359, 1983.
53. **Anon.**, Äthylenoxid mit karzinogenem Risiko, *Dtsch. Apoth. Ztg.*, 124, 1270, 1984.
54. **Frank, H. F. and Beyer, F.**, Einfluss von naturbelassenem und entkeimtem Pfeffer auf die Verderbsgeschwindigkeit verzehrsfertiger Suppen, *Dtsch. Lebensm. Rundsch.*, 80, 369, 1984.
55. **Wistreich, H. E., Thundiyil, G. J., and Juhn, H.**, Ethanol Vapor Sterilization of Natural Spices and Other Foods, U.S. Patent 3,908,031, 1975.
56. **Neumayr, L. and Leistner, L.**, *Mitteilungsbl. Bundesanstalt Fleischforsch. Kulmbach,* 72, 4600, 1981.
57. **Scharf, M. M.**, Sterilization of Spices by in situ Salt Formation, U.S. Patent 3,316,100, Ser. no. 455,327, April 25, 1967.
58. **Incze, K.**, Uj módszerek füszerek csirátlanitására (New methods in sterilization of spices) (in Hungarian), Paper presented at the Colloq. Cent. Food Res. Inst. and Food Sci. Comm. Hung. Acad. Sci., Budapest, May 1970.
59. **Huszka, T., Cséfalvy I., and Incze, K.**, Füszerpaprika sterilezése sósavval (Sterilization of powdered paprika by means of hydrochloric acid) (in Hungarian), *Konzerv. Paprikai.*, (6), 213, 1973., *Food Sci. Technol. Abstr.*, 7(2), 23246, 1975.
60. **Nakashima, K., Fukumoto, J., Nakano, N., and Ogawa, M.**, Removal of bacteria from parsley by sonication, *J. Jpn. Soc. Food Nutr.*, 29, 261, 1976; *Food Sci. Technol. Abstr.*, 10(8), T 285, 1978.
61. **Thiessen, F. M. and Hofmann, K.**, Aromatisierung: Veränderung von Gewürzen und Essenzen durch Hitzebehndlung, *Ernahrungswirtschaft,* 50, 317, 1970.
62. **Maarse, H. and Nijssen, L. M.**, Influence of heat sterilization on the organoleptic quality of spices, *Nahrung,* 24, 29, 1980.
63. **Coretti, K.**, Gewürzentkeimung, eine hygienische Notwendigkeit zur Entkeimung durch UV-Bestrahlung, *Fleischwirtschaft,* 7, 386, 1955.
64. **Walkowiak, E., Aleksandrowska, F., Wityk, A., and Watychowicz, Z. I.**, Sterilization of spices used in the meat industry by UV irradiation (in Polish), *Med. Weter.*, 27, 694, 1971; *Food Sci. Technol. Abstr.*, 6(5), S 569, 1974.
65. **Eschmann, K. H.**, Gewürze — eine Quelle bakteriologischer Infektionen, *Alimenta,* 4(3), 83, 1965.
66. **Bockemühl, J. and Wohlers, B.**, Zur Problematik der Kontamination unbehandelter Trockenprodukte der Lebensmittelindustrie mit Salmonellen, *Zentralbl. Bakteriol. Parasitenkd. Infektionskr. Hyg. Abt. 1 Orig. Reihe B,* 178, 535, 1984.
67. **Brittin, W. A.**, Spices and Herbs. CX/FH 85/9, Background paper prepared for the 21st Session of the Codex Committee of Food Hygiene, Codex Alimentarius Commission, Rome, August 1985.
68. International Commission on Microbiological Specifications for Foods (ICMSF), *Microorganisms in Foods. II. Sampling for Microbiological Analysis; Principles and Specific Applications,* University of Toronto Press, Toronto, 1974.
69. **Christensen, J. H. B.**, Review of the possible needs for microbiological specifications for codes and standards prepared or under preparation by the Codex Alimentarius Commission, Annex X, in *Rep. 2nd Joint FAO/WHO Expert Consultation on Microbiological Specifications for Foods,* EC/Microbiol./77/Rep. 2, Food and Agriculture Organization, Rome, 1977.
70. *General Code of Hygienic Practice for the Dairy Industry and Advisory Microbiological Criteria for Dried Milk, Edible Rennet Casein and Food Grade Whey Powders.* IDF Doc. 178, International Dairy Federation, Madison, Wisc., 1984.

Chapter 2

REDUCING MICROBIAL POPULATIONS IN DRY INGREDIENTS BY IONIZING RADIATIONS

I. INTRODUCTION

The application potential of ionizing radiation in food processing is based mainly on the fact that it inhibits DNA synthesis very effectively in living cells; thereby, proper dosages of ionizing radiation can sterilize and kill insects or reduce viable cell counts of microbial populations in food including dry ingredients. Research has been conducted during the past 4 decades with respect to both the technology of the process and to the suitability of irradiated commodities for food manufacturing and human comsumption.

The ionizing radiation employed in food irradiation is limited to high energy electromagnetic radiation (γ- or X-rays) with energies up to 5 MeV or electrons with energies up to 10 MeV. These radiations are chosen because:

1. They produce the desired effects with respect to the food.
2. They do not induce radioactivity in foods or packaging materials.
3. They are available in quantities and at costs that allow commercial use of the process. Other kinds of ionizing radiation, in some respect, do not suit the needs of food irradiation.

Radiation treatment causes practically no temperature rise in the product. For this reason irradiation has been termed a "cold process", leaving the food closer to its unprocessed state.

High energy irradiation can be applied through any type of packaging materials including those which cannot withstand heat. This means also that radiation can be applied to hermetically packaged products, thus avoiding recontamination or reinfestation.

Due to these characteristics, irradiation of dry ingredients is a challenging alternative to fumigation. In this regard, the largest amount of work has been devoted so far to irradiation of spices and condiments. Scientists at the Massachusetts Institute of Technology (MIT) did pioneering work in this field.[1,2] Their studies were followed up with a large variety of commodities in various countries.

II. SPICES AND CONDIMENTS

According to the International Trade Centre UNCTAD/GATT "spices may be defined as one of the various strongly flavoured or aromatic substances of vegetable origin obtained from tropical and other plants, commonly used as condiments or employed for other purposes on account of their fragrance and preservative qualities".[3] Condiments are spices alone or blends of spices which have been formulated with other flavor potentiators to enhance the flavor of foods. The so-called "true spices" (pepper, cinnamon, nutmeg, cloves, etc.) are products of tropical plants and may be bark, roots, buds, fruit, or other parts. Herbs are usually from leafy parts of plants of the temperate zone (oregano, marjoram, basil, etc.). Spice seeds (mustard, caraway, celery, anise, etc.) may be either from tropical or temperate areas. Dehydrated vegetable seasonings (onion, garlic, parsley, etc.) are also important items.

The characteristics and nomenclature of all recognized spices and condiments have been reviewed by Pruthi.[22]

Spices are widely used in food preparation, both on the domestic and industrial levels. The use of spices in the food industry had tended to increase, because of the fact that food preparation is shifting from the household to the food industry, especially for the production of the so-called "convenience foods". The acceptability of such products depends to a large extent on the seasoning blend used.

Spices are an important commodity in international trade. They are generally an expensive food ingredient, and the trade is characterized by a high value in relation to the tonnage involved. The U.S., Canada, Japan, the countries of the European Economic Community (EEC), U.S.S.R., and Switzerland are the largest importers of spices. Many spices are grown in developing countries. The largest exporters of spices in 1980 were Brazil, India, Malaysia, Indonesia, Madagascar, Mexico, China, Pakistan, Tanzania, Hungary, and Spain. Spices represent an important factor in the national economy of these countries.

A recent survey of the International Trade Centre UNCTAD/GATT on the world market of spices reported that imports into the 40 major spice-importing countries amounted to an estimated 296,000 to 327,000 ton of spices, valued at between $694 to $781 million annually during the period 1978 to 1980.[3] Imports of spices grew during the last decade, and the trend is likely to continue.

The U.S. is the largest individual market for spices in the world. Imports of spices (excluding spice herbs, mustard and sesame seeds) into this country averaged 76,000 ton annually (valued at $134 million) during the period 1976 to 1980.[3]

In industrialized countries large amounts of spices are absorbed by the industrial sector, mainly in food processing. Spices are used in virtually all categories of the food industry — meat, fish, vegetable products, bakery products, and other prepared and convenience foods. The meat industry and soup manufacturers use the largest quantities of spices. It is estimated that 45 to 50% of the spices used in the U.S. goes into manufactured food products.

Pepper is easily the leading volume and value item in the international spice trade. In recent years trade in black pepper has represented about 80% of the total. Next in importance is the capsicum group, consisting of paprika (the leading item), chilies, and cayenne pepper. Nutmeg, and mace, and cassia also feature prominently in the imports of spices into the industrialized countries of Europe and North America. Other items of significance in volume are ginger, turmeric, and spice seeds, e.g., coriander, juniper, anise, caraway, and cumin.

Dehydrated vegetable seasonings, i.e., dehydrated onions and garlic, are also important flavoring products wich have a number of advantages over the "natural product". For example, the reduction in bulk means lower transport and distribution costs, and the dehydrated products may be more readily dispersed in the food products than in the case of the chopped or blended vegetable.

It is estimated that most of the almost 500,000 ton of summer crop onions grown in California is currently used for processing.[340] The use of dehydrated onions in the EEC (15,300 ton kibbled onions plus 5100 ton powdered) in 1980 to 1981 was equivalent to over 200,000 ton fresh weight, over 10% of the total usage of fresh and stored onions. According to Heath[341] approximately 8000 ton of garlic powder is produced annually, mainly in the U.S., Europe, and the Indian subcontinent. Current dehydrated vegetable production (most of which is onion) averages about 6000 ton/annum in Egypt.[342]

International standards exist for black pepper, cardamom, ginger, chilies, and pimento (allspice), and substantial progress has been made toward setting standards for curry powder, paprika, cinnamon, cloves, turmeric, and spice seeds. However, health regulations, controls, and standards still differ from one importing country to another. Since most spice-importing countries are laying greater emphasis on protecting the consumer, health regulations are likely to be more stringently applied in the future.

Table 4
CONTAMINATION OF UNTREATED SPICES WITH AEROBIC BACTERIA[10]

Spice	No. of samples	Percentage incidence of aerobic plate count/g		
		$\geq 10^5$	$\geq 10^6$	$\geq 10^7$
Allspice	33	90	45	3
Anise	22	41	5	0
Basil	21	86	38	0
Bay	41	10	3	0
Capsicum (chili)	57	61	43	12
Caraway	17	24	6	0
Cardamom	15	40	33	0
Cassia	36	3	0	0
Cinnamon	42	25	4	2
Cloves	28	4	0	0
Coriander	30	63	26	13
Cumin	12	67	25	0
Fennel	16	26	13	0
Fenugreek	10	50	20	0
Garlic	32	37	9	0
Ginger	33	52	7	0
Mace	28	7	0	0
Marjoram	21	76	33	5
Mustard	67	10	1	0
Nutmeg	45	8	4	0
Oregano	56	32	9	0
Paprika	80	89	80	18
Pepper (black)	108	97	92	42
Pepper (white)	42	62	5	0
Sage	17	47	6	0
Savory	10	10	0	0
Thyme	19	85	53	0
Turmeric	24	96	75	29

A. Microbial Contamination of Spices

With regard to whole and ground natural spices, both the producing countries and the food industries that use spices as ingredients in various manufactured foods face the problem of spices frequently containing excessive number of microorganisms for many uses unless subjected to antimicrobial treatment.[4,5] In many cases, contaminated seasonings are responsible for the spoilage of canned meat products or may cause defective sausage products.[6-8]

Spices contain those microorganisms indigenous to the soil and plants in which they are grown and that survive the drying process. The source of contamination may be dust, insects, fecal materials from birds and rodents, and possibly the water used in some processes, e.g., soaking pepper in the preparation of white pepper. Fungi may grow on spices prior to drying or during drying, storing, and shipping.

The viable cell counts of spice samples of various origins might show deviations of several log cycles, and extremely large variations occur in the microbial count of different lots of the same spice. There is no apparent correlation between the country of origin and the quality of spice with respect to numbers of microorganisms of predominant flora.[9]

A survey of bacteriological contamination of untreated spices is shown in Table 4. A similar survey in West Germany revealed recently very similar incidences of contamination levels.[330]

Table 5
TOTAL AEROBIC BACTERIAL COLONY COUNTS AND SPORE COUNTS IN VARIOUS SPICE SAMPLES[19]

	Log counts/g at 30°C	
Spices	Total aerobic viable count	Spore count[a]
Allspice	5.8	5.9
Caraway seed	5.2	3.4
Chili	6.0	5.8
Coriander I	6.4	5.9
Coriander II	6.0	4.5
Ginger	8.4	7.9
Marjoram	6.5	4.8
Mustard	5.8	5.7
Nutmeg	5.7	5.7
Paprika I	7.0	7.1
Paprika II	6.0	5.7
Paprika III	5.4	5.4
Paprika IV	5.0	4.5
Paprika V	4.8	4.3
Pepper, black I	8.0	8.1
Pepper, black II	7.5	7.4
Pepper, black III	7.4	7.4
Pepper, white I	5.6	4.1
Pepper, white II	5.6	5.2
Pepper, white III	3.5	3.5
Mixed spices	6.3	6.2

[a] After a heat shock at 80°C for 10 min.

According to the data collated in Table 4, black pepper, turmeric, paprika, allspice, and marjoram are the spices most highly contaminated with bacteria. The aerobic plate count of these spices may sometimes reach the 80- to 100-million-per-gram level.[11,12] Anaerobes are less numerous than aerobic bacteria.[10] Thus, the use of, for example, only 0.1 to 1% of a spice containing 10^7 microorganisms per gram corresponds to an infection of 10^4 to 10^5 cell per gram of the composite food products which, of course, constitutes a significant contamination. While this may have no overt public health significance, it may contribute toward spoilage in the product in which the spice is used.

In many spices, the majority of the microbial flora consists of soil-originated aerobic mesophilic spore-forming bacteria, e.g., *Bacillus subtilis, B. licheniformis, B. megaterium, B. pumilus, B. brevis, B. polymixa,* and *B. cereus.*[7,9,13-18]

As shown in Table 5, aerobic bacterial spores may frequently form more than 50% of the mesophilic total viable cell count. This is understandable since during drying and storage many vegetative cells die off while spores survive easily. The proportion of obligate anaerobe sporeformers is usually small.[17,20] Thermophilic anaerobes and aerobes are found occasionally, sometimes in moderate numbers.[21,22] Sheneman[23] found that the bacterial flora of dried onions consists mainly of Bacillus species, i.e., *B. subtilis, B. licheniformis, B. cereus,* and *B. firmus.*

Psychrotrophic or psychrophilic spore-formers are missing in spices and herbs even in those of high mesophilic counts.[24] Apparently, psychrotrophic cells (growth at 7°C) are less numerous in spices and herbs than mesophilic ones. de Boer and Janssen[25] found that the psychrotrophic counts was less than 10^5/g in 88% of 143 samples and less than 10^3/g in

Table 6
CONTAMINATION OF UNTREATED SPICES WITH COLIFORMS AND *ESCHERICHIA COLI*[10]

Organisms	No. of samples	% incidence of viable counts/g $\geq 10^2$	$\geq 10^3$	$\geq 10^4$	$\geq 10^5$
Coliform	228	25	16	7	3
E. coli	228	6	2	0	0

53% of samples. Higher than 10^6/g psychrotrophic colony-forming units (CFU) were found in some samples of thyme, dill, coriander, basil, chevril, and licorice.

Powers et al.[26] tested 110 samples of various spices for incidence and levels of *B. cereus*. The organism was found in 53% of the spices, and counts ranged from 50 to 8500/g. This incidence was more than double the incidence found in other selected dry products.[27] Of the isolates, 89% were enterotoxigenic. The frequent occurrence of *B. cereus* in spices was also reported by other authors.[25,28] In extreme cases *B. cereus* counts up to 10^5/g were also found.[29]

A relatively high incidence of *Clostridium perfringens* is also found.[25,27,30-34] de Boer and Boet[308] recently found *C. perfringens* in 80% of 54 different kind of spices. Since spores of these organisms may survive cooking temperatures and will grow in foods at temperatures between 20° to 50°C, spices harboring these spores must be considered as a potential health hazard. *B. subtilis,* common in many spices, has been in a few cases also held responsible for food-borne gastroenteritis.

A wide variety of nonsporing bacteria may also be present in spices.[9,35] Coliforms are often found, but *Escherichia coli* is infrequent (Table 6).[10,29,36-38] de Boer and Janssen[25] found that 23% of 153 samples of spices and herbs contained more than 10^4 Enterobacteriaceae per gram.

Spices may not be suitable substrates for the growth or long survival of Salmonellae, which are usually absent from them.[9,33,39,40] However, it is not certain if, in some cases, this is not due to inadequate methodology (lack of resuscitation, inhibition by spice components during counting procedure).[10,41] Therefore, occasional *Salmonella* contamination in 6.7% of spices and herbs samples examined (*S. glostrup, S. heidelberg, S. schwarzengrund,* and *S. typhimurium* have been found in some samples of pepper, coriander, peppermint, and paprika, respectively). Indeed, black and white pepper have both been implicated as vehicles for the spread of *S. weltevreden* resulting in several serious cases of salmonellosis,[43-45] while black pepper proved also to be the source of a *S. oranienburg* epidemy in 1981 and in 1982 in Norway in which over 120 patients were recorded.[46,319]

Fecal streptococci occur in about one half of the spice samples, usually in low number, rarely reaching the level of 10^4/g.[29,47] Staphylococci and lactic acid bacteria are very rare in spices.[25,29,47]

According to several authors, spices play a major etiological part in mold contamination of meat products.[5,39,48] Mold counts of spices do not correlate with the aerobic plate counts.[18] White pepper, black pepper, chili, and coriander seem to be most heavily contaminated with molds (see Table 7).[10]

Although components of the mold flora may vary greatly with the spice, the *Aspergillus glaucus* group, *A. niger, Penicillium* spp. are usually most prevalent.[39,48-54] Vaughn[55] found that *Mucor, Penicillium,* and *Aspergillus* are the main fungi in dried onions.[56] Eckhardt and Leistner[56] found *A. flavus, A. niger, A tamarii, A. fumigatus,* and *A. nidulans*. Ito et al.[18] found mainly *A. glaucus* group, *A. restrictus* group, *A. flavus* group, *A. fumigatus, A. niger,* and *Penicillium*.

Table 7
CONTAMINATION OF UNTREATED SPICES WITH MOLDS[10]

Spice	No. of samples	Percentage incidence of mold count/g		
		$\geq 10^4$	$\geq 10^5$	$\geq 10^6$
Allspice	27	25	7	0
Anise	16	13	0	0
Basil	17	6	0	0
Bay	35	11	0	0
Chili	59	19	12	5
Caraway	14	7	0	0
Cardamom	15	0	0	0
Cassia	20	15	10	0
Cinnamon	51	6	0	0
Cloves	26	4	0	0
Coriander	23	35	9	0
Cumin	8	13	13	0
Fennel	11	0	0	0
Fenugreek	8	25	25	0
Garlic	15	0	0	0
Ginger	28	11	0	0
Mace	22	4	0	0
Marjoram	14	29	0	0
Mustard	63	2	0	0
Nutmeg	33	9	0	0
Oregano	48	9	0	0
Paprika	61	5	0	0
Pepper (black)	82	30	25	23
Pepper (white)	44	61	25	2
Sage	14	50	0	0
Savory	6	0	0	0
Thyme	16	87	6	0
Turmeric	32	3	0	0

Relatively high incidence of toxigenic molds has been detected.[39,51,55,57,58] Several authors have found aflatoxins in a range of spices (black pepper, ginger, turmeric, celery seed, nutmeg) although the levels of aflatoxins recorded were generally low.[59-61,309] While certain spices and herbs, especially cinnamon, cloves, and possibly oregano and mustard, inhibit mycelial growth and subsequent toxin production, others, particularly sesame seed, ginger, and rosemary leaves, appear to be conducive to aflatoxin occurrence.[62] A survey from France implicated pepper as the source for the toxigenic *A. flavus* and for high levels of aflatoxin in sausages and pepper cheese.[338]

Yeasts have been found in low numbers only.[29,47]

B. Limitations of the Use of Volatile Oils and Oleoresins

As alternatives for natural spices volatile oils obtained by steam distillation or oleoresins obtained by extraction with organic solvents offer certain advantages over natural spice powders, i.e., consistency of their quality, freedom from enzymes and microorganisms, uniform dispersion in the products, and easy handling and storage.[7,63,67] These spice extracts may be mixed with a carrier, e.g., salt, or encapsulated, e.g., in starch.

Meat packers in several countries however, do not prefer such "spice aromas" to whole spices. This is because they are unable to substitute all functional properties of the original spices, e.g., aroma profile or appearance of the product. In products such as salami, sausages, etc. they make a point of showing the spices, particularly pepper and cloves. The "bolder" the pepper so exhibited, the greater apparently the appreciation of the spice quality of the meat products displayed. Another problem in the use of spice extracts is that their flavor quality generally does not equal that of the whole or ground spices.[68,69] Comparing the spicing power of ground pepper and pepper oil in bologna sausage, Körmendy[69] found that by using ground pepper a more intense taste is obtained even if the concentrations of the spicing agents were equal according to the flavor profilograms of their dilutions. The reason for this fact is that in contrast to the homogeneous distribution of pepper extract in the sausage, ground pepper is inhomogeneously distributed, forming "hot spots" in the product. When these "spots" are crushed in the mouth, they release a pungent taste sensation. Due to their volatility, the loss of flavor of spice extracts during any cooking process, unless fully enclosed as in a can, is greater than that of natural spice.[63] Another disadvantage of oleoresins arises from the need for a solvent in their production. Unless great care is exercised in ensuring the purity of the solvents, high boiling trace of solvents may well remain. These are of concern because of eventual unpleasant flavor and toxicity. Sometimes spices that are too moldy for use as ground spice are extracted when the solvents can also extract mycotoxins.[10] The result of Danish studies of commercial spice extracts used in meat products and in neutral sauces indicated large variations in strength and flavor of extracts, even from the same manufacturer, compared to the natural spices from which they were extracted.[315] Liquid CO_2 extraction of spices has been tested and found to be better than extraction with commonly used solvents, i.e., trichloroethylene.[70] The method is commercially not yet in operation.

C. Radiation Dose Requirement for Cell-Count Reduction in Spices and Herbs

In radiation processing the most important parameter is the absorbed radiation dose required to obtain the specific desired radiation effect.

A survey of microbiological studies at various radiation doses on the total mesophilic cell counts is given in Table 8 as compared to microbiological studies reported on ethylene oxide (EtO) treatment.

It can be seen from Table 8 that as with any antimicrobial treatment, the number of surviving microorganisms depends on the initial level of viable cell counts. Depending on the number and types of microorganisms and the chemical composition of the spice, a radiation dose of up to 16 to 20 kGy may be required to achieve "sterility" (i.e., a reduction of the total viable cell count to less than 10/g) in natural spices. However, already doses of 3 to 10 kGy can reduce the total aerobic viable cell counts below a level of 10^3 to 10^4/g. This level is generally required and accepted as a maximal count of decontaminated spices (in German: "entkeimte Gewürze") demanded in the spice trade.[7,11,71,72] The dose range of 3 to 10 kGy is approximately equal in microbicidal effect to the commercially established fumigation process.

Surveying the large number of publications referred to in Table 8, one can also conclude that there is not much difference in the radiation resistance of the aerobic spores most frequently occurring in spices (with an apparent overall D-value varying between less than 1.7 and 2.7 kGy), and this is not much affected, at least from the practical point of view, by the water activity of their environment. In fact, the D-values which could be derived from irradiation of spice samples are similar to those found for related pure strains of aerobic spore-formers in aqueous systems.[78,79] At the same time, it is known that their resistances against heat or chemicals are widely differing and very much influenced by the environment. This reflects the differing modes of action of various sporodical agents.

It is apparent from Table 8 that in several cases (e.g., cayenne, cumin, garlic powder, and peppers) EtO treatment did not always reduce the viable cell counts below the 10^4/g level. Data collated by the International Commission on Microbiological Specifications for

Table 8
COMPARISON OF EFFECTS OF IRRADIATION AND EtO TREATMENT ON TOTAL VIABLE CELL COUNT (TAVC) OF VARIOUS SPICES AND CONDIMENTS

	log TAVC/g																						
		Irradiated at (kGy)																					
Spice	Untreated	1	2	2.5	3	4	4.5	5	5.5	6	6.5	7.5	8	9	10	11	12	15	16	20	EtO treatment	Fumigation conditions	Ref.
Allspice	6.8																				<1.0	500 g ETOX/m³, 6-7 hr, 20-25°C	81
	6.6																				1.6	1000 g ETOX/m³, 6 hr	82
	6.4														<1.0								83
	6.2		5.0			4.0		4.0		3.0		1.9	1.3		<1.0						1.6		84
	5.7							2.3							1.0								15
	5.0																				0.4		85
Anise	5.3							2.2															86
Basil	6.5														3.0								83
	6.0					4.0																	85
	4.4																			<1.0			87
Bay	5.6												1.9								2.6		85
	5.5							2.0							<1.0								15
	5.3												1.8								2.4		85
Caraway	6.3							<1.0				<1.0			<1.0								88
	5.8											2.0											89
	5.6																				<1.0		81
	5.5							5.0				3.6			2.3								73
	4.5																				3.0	250 g/m³, 22°C, 6 hr	90
	4.5																				2.0	750 g/m³, 22°C, 9 hr	90
	4.3																				2.2	500 g/m³, 22°C, 6 hr	90
	4.3																				2.0	500 g/m³, 22°C, 9 hr	90
	4.3																				2.0	500 g/m³, 22°C, 18 hr	90
Cardamom	6.3																				3.6	250 g/m³, 22°C, 6 hr	90
	6.3																				3.1	750 g/m³, 22°C, 6 hr	90
	6.1												<2.7								<2.0		89
	5.8																				<1.0		81
	5.7												<1.0								2.4		85
	4.1							2.5				1.4											91
Cayenne	7.5				5.5		3.5			2.3		1.8			<1.0								92
	7.0												2.6								5.7		85
	6.0																				1.8	500 g ETOX/m³, 20-25°C, 6-7 hr	81
Celery leaves	4.1					2.8															3.3		85

Celery seeds	5.6								<1.0						83
	5.6	4.8	3.8	4.0		3.0	2.2	1.8	<1.0				0.9	16 hr	84
Charlock	4.9			3.6			2.0								73
Chive	6.1		3.3										3.4		85
	6.0	3.8	3.6-3.9					<3.0-3.8	3.6						93
	5.9							<3.2							93
Cinnamon	5.8			3.3					<2.0						94
	4.2	3.7	3.4			<2.0							<1.0	1500 g/m^3, 20-25°C, 8 hr	307
	3.3			2.1			<1.0								95
Clove	5.5												1.0	500 g/m^3, 20-25°C, 6-7 hr	81
	3.9			2.3											86
	3.5							<2.0					<2.0		89
	3.3											<1.0			87
	2.9			1.6			<1.0								91
	2.9			1.6			<1.0								95
Coriander	7.4			3.5			<1.0		<1.0						88
	6.7			2.9					<1.0						94
	6.4												2.7	500 g ETOX/m^3, 20-25°C, 6-7 hr	81
	6.2			2.4											86
	6.0								2.3						96
	5.8							2.2					1.8		85
	5.7							1.5					2.0		85
	5.3		3.7					1.8			<1.0				19
	5.3												3.0	500 g/m^3, 22°C, 6 hr	90
	5.3												2.4	500 g/m^3, 22°C, 9 hr	90
	5.3												2.0	500 g/m^3, 22°C, 18 hr	90
	5.0												3.2	250 g/m^3, 22°C, 6 hr	90
	5.0												2.0	250 g/m^3, 22°C, 9 hr	90
Cumin	7.0							2.4					4.5		85
	6.7							2.7							85
	6.3			2.0											86
	4.0-6.1	3.3-4.6	2.0-3.6			<2.0-3.0		<2.0-2.0	<2.0						97
	5.4			4.0					2.5	<1.0					15
Curry	7.9			5.4				3.5	2.8						96
	7.3		4.3			2.6		2.0							96
	>7.0								2.8						96
	6.3							4.1					3.6		85
	6.3				5.3										96
	6.3								<2.0						96
Dilltips	6.0			4.8			3.7		3.6	3.3					98, 99
Fennel	5.6	4.8	4.2			3.9		3.2	2.7				2.1	1500 g/m^3, 20-25°C, 8 hr	307
	5.5												1.6	75 oz EtO/cuft, 125°F, 4 hr	100

Table 8 (continued)
COMPARISON OF EFFECTS OF IRRADIATION AND EtO TREATMENT ON TOTAL VIABLE CELL COUNT (TAVC) OF VARIOUS SPICES AND CONDIMENTS

		log TAVC/g																					
		Irradiated at (kGy)																					
Spice	Untreated	1	2	2.5	3	4	4.5	5	5.5	6	6.5	7.5	8	9	10	11	12	15	16	20	EtO treat-ment	Fumigation conditions	Ref.
Fenugreek	6.8							4.9							4.1								94
	4.9		3.6			3.4				2.3											<1.0	1500 g/m^3, 20-25°C, 8 hr	307
Garlic powder	5.9					<1.0							<1.0						<1.0		5.3	800 g/m^3, 22°C, 6 hr	19
	5.9					4.5							2.1										320
	5.7							3.8				3.0			2.3			2.0					98, 99
	4.7							1.3				0.5									4.2	5 hr	84
Ginger	6.8		5.6			4.1				3.6											3.1	1500 g/m^3, 20-25°C, 8 hr	307
	6.7																				4.1	1000 g ETOX/m^3, 6 hr	82
	6.6												2.4								3.7		85
	6.3			5.8				3.4															101
	6.0																				3.8		102
	5.6														2.9								83
	5.2																				<1.0	500 g ETOX/m^3, 20-25°C, 6-7 hr	81
	3.0							1.0							<1.0								15
Juniper	6.0											4.0			2.3								73
Lemon peel	4.9							3.7							3.0								94
Mace	4.6							2.7				<1.0											91
	4.4																				1.8	1000 g ETOX/m^3, 6 hr	82
Marjoram	6.9																				1.7	500 g ETOX/m^3, 20-25°C, 6-7 hr	81
	6.8					2.5							<1.0					<1.0					19
	6.5							4.8							3.0								94
	5.7							4.5				3.0			2.7			2.0					98, 99
	4.0																				<1.0		103
Mustard seed	5.3							2.7				2.7			2.3			2.0					98, 99
	5.3					1.8																	85
	2.7																				<1.0	500 g ETOX/m^3, 20-25°C, 6-7 hr	81

Nutmeg	6.4					2.8											104
	6.0					<2.0				<2.0							89
	5.8			4.2		2.7											101
	5.7														1.5		81
	5.5				3.6					1.6					2.2		85
	5.4					2.0						2.0					94
	5.0					2.1											104
	4.8					2.5											105
	4.6					<1.0											106
	4.3														1.3	1300 g ETOX/m^3, 6 hr	82
	4.0					1.5											104
Onion powder	7.6									4.5							85
	7.2				5.4					4.9		4.1					107
	5.9				3.8					2.6			<1.0				19
	5.7											3.4					83
	5.7										2.6						107
	5.3											2.2					96
	5.2				3.6					3.1					4.6	800 g/m^3, 22°C, 6 hr	320
	4.9										4.6						107
	4.7					1.9						1.5					108
	4.5										2.8						107
	4.5				2.3	1.8						1.3					109
	3.8	2.8	2.5							1.6				<1.0			110
Orange peels	4.0					4.0						2.5					110
Oregano	6.5		2.4		1.6		<1.0										84
	6.2											1.5					83
	6.1											<1.0					83
	5.5				2.9										2.5		85
	5.5				3.0										2.0		85
	4.5					<1.0									<1.0	16 hr	84
Paprika	7.2-7.7		6.2-6.8		5.1-5.9		4.3-5.4			3.2-3.9		<2.0-3.1					97
	7.0-7.9		6.0-7.8		4.4-5.7		3.0-4.9			2.0-4.6		<2.0					97
	7.1					4.8		4.0				2.2	2.0				98, 99
	7.0		6.0		5.0	3.9	3.8	3.1		2.7		<1.0			<1.0	16 hr	84
	6.8		3.5		2.5												112
	6.8														<1.0	500 g ETOX/m^3, 20-25°C, 6-7 hr	81
	6.7					3.5				1.9		<1.0					113
	6.6			5.0	4.0												114
	6.3							2.0									83
	6.2														4.3	1000 g ETOX/m^3, 6 hr	82
	6.2					<1.0			<1.0			<1.0					88
	5.7					3.4											108
	5.7					3.0											115

Table 8 (continued)
COMPARISON OF EFFECTS OF IRRADIATION AND EtO TREATMENT ON TOTAL VIABLE CELL COUNT (TAVC) OF VARIOUS SPICES AND CONDIMENTS

	log TAVC/g																						
		Irradiated at (kGy)																					
Spice	Untreated	1	2	2.5	3	4	4.5	5	5.5	6	6.5	7.5	8	9	10	11	12	15	16	20	EtO treat-ment	Fumigation conditions	Ref.
	5.5					2.2							<1.0						<1.0				19
	5.3												1.0								2.2		85
	5.0							2.7						2.2	1.3						2.4-3.6		116
	5.0											2.7			2.3								98
	4.9							2.6															108
	4.9							2.4						2.1		1.5					3.0	600 g ETOX/m^3, 6 hr	109
	4.7					2.4							1.4								2.0	800 g/m^3, 22°C, 6 hr	320
	4.5			3.5				2.6															101
Parsley	6.9					3.8															3.6		85
	6.8												2.5								3.5		85
	5.7							3.0						2.8				2.3					98, 99
	4.6-5.2		4.5-4.6																				93
	3.3					<1.0							<1.0										19
Pepper, black	8.0																				3.0		81
	8.0		6.2			5.2				3.9			2.1		<1.8								12
	7.9							5.2				4.2			3.7			1.6					98, 99
	7.6			5.8				4.7							3.5								117
	7.6							3.8															86
	7.6					4.6									1.8				<1.0				19
	7.5														1.8								83
	7.5																				4.1	250 g/m^3, 6 hr	90
	7.5																				3.5	750 g/m^3, 6 hr	90
	7.4							4.0															118
	7.3												2.7								3.9		85
	7.3																				4.0	250 g/m^3, 6 hr	90
	7.3																				3.4	750 g/m^3, 6 hr	90
	7.2							3.1				2.6											91
	7.2							4.1							2.6								117
	7.0			4.8				3.0							<2.0								119
	7.0												2.4										85
	7.0							3.5							2.0			<1.0					15
	6.1-7.0		4.0- >7.0			<2.4-4.6				<2.0-2.0			<2.0		<2.0								97
	6.9			5.9				5.5							4.0								119

	6.8			5.6		3.9											117
	6.8				4.5					2.4					5.3	800 g/m³, 22°C, 6 hr	320
	6.7					3.7				2.7		1.6					113
	6.6		5.8		4.8	4.0	4.0		3.3	2.9	2.5		1.6		3.2	16 hr	84
	6.6			4.6	3.5												109
	6.6														4.3		102
	6.6														3.9	500 g/m³, 6 hr	90
	6.6														3.0	500 g/m³, 9 hr	90
	6.6														2.4	500 g/m³, 18 hr	90
	6.5			4.5											4.0	600 g ETOX/m³, 6 hr	75
	6.4			4.6		2.7											113
	6.4					2.8											109
	6.4					4.0											77
	6.3			4.9		3.9											101
Pepper, white	6.5														2.5		81
	6.3		5.7		4.9		3.7			3.2	1.8						18
	5.8										<2.0						96
	5.8					2.7								<1.0			106
	5.6					4.2									0.3		85
	5.5									4.8					5.0		85
	5.4														2.9	250 g/m³, 6 hr	90
	5.4														2.0	750 g/m³, 6 hr	90
	5.3					2.9											118
	5.2					4.5											104
	5.2					2.1											104
	4.7					2.6											105
	4.4														2.1	500 g/m³, 6 hr	90
	4.4														2.0	500 g/m³, 9 hr	90
	4.4														2.0	500 g/m³, 18 hr	90
	4.3					2.7											105
	4.3					1.2											86
Red pepper	6.6														4.3		102
	6.0					2.8					<1.0		<1.0	<1.0			120
	5.1							<1.0									83
Sage	5.8				3.5										3.3		85
	4.0														2.7		102
	3.8			2.5		<2.0											119
Savory	3.3					<1.0											15
Thyme	7.0					4.0					2.9						94
	6.4					2.9				<2.0							89
	6.3					3.5			2.8		<1.0						88
	5.3									1.3					3.1		85
	5.2										1.6						83
Turmeric	7.9					4.9											94
	7.6	6.9	6.3		4.3		3.8			2.4	1.8						18
	6.7					2.0					<1.0						15
	6.4														2.5	1000 g ETOX/m³, 6 hr	82

Table 8 (continued)
COMPARISON OF EFFECTS OF IRRADIATION AND EtO TREATMENT ON TOTAL VIABLE CELL COUNT (TAVC) OF VARIOUS SPICES AND CONDIMENTS

	log TAVC/g																						
		Irradiated at (kGy)																					
Spice	**Untreated**	**1**	**2**	**2.5**	**3**	**4**	**4.5**	**5**	**5.5**	**6**	**6.5**	**7.5**	**8**	**9**	**10**	**11**	**12**	**15**	**16**	**20**	**EtO treat-ment**	**Fumigation conditions**	**Ref.**
Mixed seasonings	6.6							3.3															120
	6.3																				5.2	400-600 g T/m^3, 20-25°C, 4-6 hr	121
	6.0				3.1																		74
	6.0							3.5															108
	5.1																				2.7		102
	5.0					3.7																	85
	4.3							2.6															109
Green tea	4.9																			<1.0			87
Oolong tea	3.5																			<1.0			87
Orange pe-koe tea	4.5																			<1.0			87

Foods (ICMSF) show that bacteria in mustard and granulated onion also have low susceptibility to destruction by EtO.[10] A survey of "decontaminated" (most probably EtO treated) spices commercially available in Germany has revealed that the total aerobic viable cell counts in 30 samples varied between 150 and 6,960,000, while the counts of mold propagules ranged from 150 to 1,090,000.[72]

A comparison on the mold count-reducing effect of irradiation and fumigation is shown in Table 9, which reveals that a dose of 4 to 5 kGy can eliminate molds at least as efficiently as EtO treatment.

For the canning industry, important components of the microflora of spices are the thermophilic bacteria. As Table 10 illustrates, they can be practically eliminated with the same radiation doses as those necessary for a sufficient reduction of the total aerobic viable cell counts.

Bacteria of the Enterobacteriaceae family are relatively radiation sensitive even in dry ingredients, and in most cases a dose of approximately 5 kGy seems to be sufficient for their elimination (Table 11).

Sulfite-reducing clostridia, usually present in low ($<10^3$/g) number, can be eliminated by 4 kGy.[19]

Table 12 is one example of detailed studies on the effect of irradiation on the microbial community structure of spices and clearly shows the efficacy of the radiation treatment of black pepper.[12]

It can be seen from Table 12 that a radiation dose of 6 kGy resulted in 4 $\log_{10}$ cycles reduction of the aerobic mesophilic colony count. The aerobic mesophilic bacterial spore count, surviving a heat treatment of 1 min at 80°C and the heat-resistant fraction of the aerobic mesophilic bacterial spores surviving a heat treatment of 20 min at 100°C, was reduced by 3 and 6 $\log_{10}$ cycles, respectively, at a dose of 4 kGy. Molds were reduced from 4.1×10^4 to below 50 CFU/g by a dose of 4 kGy.

No postirradiation recovery of surviving microorganisms was noted during storage of irradiated spice samples; on the contrary, a further decrease of survivors was reported in some cases.[73]

The germicidal efficiency of irradiation is much less dependent on the moisture conditions than the efficiency of EtO treatment.[74,75,80]

Comparative investigation on the effects of γ-radiation and EtO treatment on the bacterial spore flora of black pepper showed that in the water activity range of $a_w = 0.25$ to 0.75 studied, the water activity of the spice did not influence notably radiation sensitivity of the bacterial flora while the efficacy of fumigation was much higher at $a_w = 0.75$ and 0.50, respectively, than at $a_w = 0.25$.[76,77]

D. Effect of Irradiation on Chemical Constituents of Spices and Herbs

The chemical composition of spices is very complex. Their flavor and color characteristics are due to components which vary substantially among spices. The aroma is derived from multicomponent essential oils.

No substantial changes were found in the volatile oil content by steam distillation in most spices when they were treated with doses up to 10 to 15 kGy. A survey of such studies is given in Table 13. Interpreting these data, one should keep in mind that the variation coefficients of the volatile oil estimation are found in the range of 4 to 11%.[123]

Canadian and Hungarian authors found less damage to the volatile oil content of allspice and black pepper decontaminated by gamma irradiation than in the same spices treated with EtO.[130,320] Similarly, nonvolatile oil content of these spices plus celery seeds, oregano, and garlic and color of paprika were more affected by EtO than by gamma irradiation.[84]

Numerous authors have carried out more detailed chemical studies in connection with the radiation decontamination of spices. Gas chromatographic studies on the qualitative and

Table 9
COMPARISON OF THE EFFECTS OF IRRADIATION AND EtO TREATMENT ON MOLD COUNTS OF VARIOUS SPICES AND CONDIMENTS

Spice	Untreated	Log mold count/g irradiated at (kGy)											EtO treatment	Fumigation conditions	Ref.
		1	2	2.5	3	4	4.5	5	6	7.5	8	10			
Basil	2.6											<1.0			83
Caraway	3.0							<1.0							88
Cardamom	3.1	1.7						<1.0							91
Cayenne	5.3				3.8		3.0		2.0	<1.0		<1.0			92
Celery seeds	2.3											<1.0			83
Cinnamon	5.3							1.0				1.0			94
	3.7		<1.0			<1.0			<1.0				<1.0	1500 g/m³ 20°-25°C, 8 hr	307
	2.5	1.0													91
Clove	3.0	1.6						<1.0							91
Coriander	5.7							<1.0							88
	4.3					<1.0						<1.0			94
	4.2					<1.0									19
Cumin	4.9							<1.0							86
Dill tips	2.5							2.3		2.0		<2.0			99
Fennel	3.0		2.2			<2.0			<1.0				1.8	1500 g/m³ 20°-25°C, 8 hr	307
Fenugreek	3.4							1.3				1.6			94
	2.5		1.6			<1.0			<1.0				<1.0	1500 g/m³, 20°-25°C, 8 hr	307
Garlic powder	3.9											<1.0			83
	2.8					1.5					1.0		2.2	800/m³ 22°C, 6 hr	320
	2.5							<2.0							99
	<1.0					<1.0									19
Ginger	3.5		2.2			<1.2			<1.0				<1.0	1500 g/m³ 20°-25°C 8 hr	307
	3.2												<1.0		102

Lemon peel (powdered)	4.9					<1.0				<1.0			94
Mace	3.9	1.9				<1.0							91
Marjoram	4.3				<1.0								19
	4.2					<1.0							94
	3.6					<2.0							99
Mustard seed	2.8					2.0		<2.0					99
Nutmeg	4.0					1.0				1.0			94
Onion powder	4.8		3.4		2.7		2.5		2.5				122
	4.2				1.8				0.7		2.4	800 g/m³ 22°C, 6 hr	320
	3.4	2.2	0.8										110
	3.0				<1.0								19
	2.0					<2.0							99
Orange peel	3.0					1.0				1.0			94
Oregano	4.0									<1.0			94
	3.3		2.2		<1.0								84
Paprika	5.7					<1.0							88
	3.6					1.2					2.2		111
	2.7					<2.0							99
	2.7					<1.0					2.6	600 g/m³ 25°C, 6 hr	109
	2.3				<1.0								19
	2.2				1.2				<1.0		1.7	800 g/m³ 22°C, 6 hr	320
Parsley	2.3					<2.0							99
Pepper, black	4.6			<1.0									117
	4.8					<2.0							99
	4.4				<1.0				<1.0		<1.0	800 g/m³ 22 °C, 6hr	320
	3.7				<1.0								19
	3.2										<1.0		102
	2.6	1.1				<1.0							91
Pepper, white	3.1										<1.0		102
Sage	2.8										<1.0		102
Thyme	5.1					<1.0				<1.0			94
	3.5									<1.0			83
	2.0					<1.0							88
Turmeric	2.7					<1.0				1.0			94
Mixed seasoning	2.7										<1.0		102

Table 10
COMPARISON OF THE EFFECTS OF IRRADIATION AND EtO TREATMENT ON THERMOPHILIC AEROBIC BACTERIAL COUNTS OF VARIOUS SPICES AND CONDIMENTS

	Log counts of thermophilic aerobic bacteria/g											
		Irradiated at (kGy)										
Spice	Untreated	2	2.5	4	5	6	8	10	16	EtO treatment	Fumigation conditions	Ref.
Allspice	6.2	4.8		3.5		2.0	0.6					84
Celery seeds	5.1	3.9		2.5		1.8	<1.0					84
Coriander	5.0			2.8			1.5		<1.0			19
Garlic powder	4.4			<1.0			<1.0		<1.0			19
	4.1			3.3			2.0			3.8	800 g/m^3, 6 hr	320
	3.0						<1.0					84
Marjoram	<4.0			<1.0			<1.0		<1.0			19
Onion flakes	5.9			3.5			2.2		<1.0			19
Onion powder	2.8			<1.0			<1.0			<1.0	800 g/m^3, 6 hr	320
Paprika	5.5	4.5		2.3		<1.0						84
	5.3			1.8			<1.0		<1.0			10
	3.2			1.0			<1.0			1.9	800 g/^{3}, 6 hr	320
Parsley	2.6			<1.0			<1.0		<1.0			19
Pepper, black	7.5			4.0			1.5		<1.0			19
	7.3		4.5		<1.5			<1.5				119
	6.6		5.7		4.4			3.6				119
	6.2	5.0		3.7		2.5	1.7	<1.0				84
	5.5			2.9			1.2			3.2	800 g /m,^{3} 6 hr	320
Sage	2.7				<1.5							119

Table 11
EFFECTS OF IRRADIATION AND EtO TREATMENT OF ENTEROBACTERIACEAE COUNTS OF VARIOUS SPICES AND CONDIMENTS

		Log Enterobacteriaceae counts/g									
		Irradiated at (kGy)									
Spice	Untreated	2	2.5	4	5	6	8	10	Eto treatment	Fumigation conditions	Ref.
Cinnamon	2.0				1.0			1.0			94
Coriander	5.1				<1.0			<1.0			94
Fennel	5.0	4.1		3.9		3.5	2.7	<2.0	<1.0	1500 g/m^3, 20°-25°C, 8 hr	307
Fenugreek	5.4				<3.0			1.6			94
	2.8	<2.0		<1.0		<1.0			<1.0	1500 g/m^3, 20°-25°C, 8 hr	307
Ginger	4.6	3.7		2.5		<1.0			<1.0	1500 g/m^3, 20°-25°C,8 hr	307
Lemon peel	3.2				1.5			<1.0			94
Marjoram	3.3				<1.0			<1.0			94
Nutmeg	2.0				1.0			<1.0			94
Onion powder	3.8			<1.0			<1.0		1.3	800 g/m^3, 22°C, 6 hr	320
Orange peel	<2.0				<1.0			<1.0			94
Paprika	2.7			<1.0			<1.0		<1.0	800 g/m^3, 22°C, 6 hr	320
Pepper, black	5.3		4.1		2.7						94
	3.8				2.2						86
	3.4		2.4		<1.0						94
	3.2			1.1			<1.0		<1.0	800 g/m^3, 22°C,6 hr	320
	1.8		<1.0								94
Thyme	5.1				1.8			1.0			94
Turmeric	4.4				1.0			1.0			94

Table 12
MICROBIAL DECONTAMINATION OF BLACK PEPPER BY γ-RADIATION[12]

Microorganism	Logarithm of CFU at (kGy) 0	2	4	6	8	10
Total aerobic mesophilic	8.0	6.2	5.2	3.9	2.1	<1.8
Aerobic mesophilic (spores)						
Surviving 1 min at 80°C	7.7	6.6	4.7	3.0	1.8	<1.8
Surviving 20 min at 100°C	6.0	2.9	0.2	—	—	—
Anaerobic mesophilic (spores)						
Surviving 1 min at 80°C	7.5	6.1	3.1	<1.8	<1.8	<1.8
Surviving 20 min at 100°C	5.9	<2.8	<1.8	<1.8	<1.8	<1.8
Enterobacteriaceae	4.7	2.8	1.7	1.1	<−0.5	—
Lancefield D streptococci	4.9	1.7	0.4	<−0.5	<−0.5	—
Molds	4.6	<1.8	—	—	—	—

quantitative composition of volatile oils from irradiated allspice, black pepper, cumin, cloves, white pepper, caraway, cardamom, nutmeg, mace, and coriander showed that the percentage distribution of components of volatile oils remained almost unchanged.[91,123-125,131]

As examples, the quantitative results of gas chromatographic analyses of essential oils in irradiated and untreated cardamom and white pepper are shown in Tables 14 and 15, respectively.[96,125] Nevertheless, slight quantitative changes may occur in some cases. Bachman and Gieszczynska[73] found that the content of terpenes, i.e., α-terpinene, γ-terpinene, and terpineole, decreased in the volatile oil of irradiated marjoram, and the decrease in terpene content was accompanied by an increase in the alcoholic fraction of the essential oil. α-Pinene and Δ^3-carene appeared to be resistant to irradiation. Considerable quantitative differences in the percentages of a number of constituents of volatile oil were reported by Theivendirarajah and Jajewardene[91] between control and irradiated (10 kGy) samples of cinnamon bark, as shown in Table 16.

The yield of lipids by Soxhlet extraction tended to increase in irradiated cardamom seeds.[125] However, the fatty acid composition of the lipid fraction did not change up to 50 kGy (Table 17). No significant changes in total and reducing sugars and pentoses were noted in the dose range of 5 to 50 kGy, either.[125]

The lipid content of caraway seed by Soxhlet extraction as well as its chemical indexes are shown in Table 18 as a function of the radiation dose. The slight increase in the apparent lipid content can be noted here as well. The acid value and the peroxide value increased slightly, while the iodine value decreased slightly at doses over 10 to 15 kGy.

A slight increase in the total and reducing sugar contents of caraway seeds as noted with increasing doses, while the desoxy sugar fraction remained practically unchanged up to 60 kGy.[125]

In nutmeg an increase of methanol extractable substances and a decreased proportion of $C_{18:2}$ fatty acid in the methanolic extract were found after irradiation at 5 and 15 kGy.[312]

There was no decrease in piperettine and piperine contents of black and white peppers with respect to γ-radiation, even up to 45 to 50 kGy.[88,117,123,131,132]

Capsaicin, the pungent principle of hot paprika, has been found at a level of 34.7 mg% in unirradiated paprika and a level of 34.0 mg% in samples irradiated at 16 kGy.[74] The difference is within the error range of the analytical method.

Some increase of the yield of ether- or petroleum ether-extractable fraction was noted in paprika irradiated at the dose range of 5 to 15 kGy.[130,132]

However, neither the vitamin C nor the linoleic acid and linolenic acid content of paprika was affected to a significant extent by doses of up to 16 kGy.[74,312] No decrease in benzene-soluble pigment content (carotenoid content) and reducing sugar content was observed following treatment with doses up to 4 kGy, and 90% retention of pigment was found after treatment with 16 kGy.[74,80] Neumayr and co-workers[19] did not find any significant changes in the hexane-soluble pigment content of paprika up to 16-kGy dose level studied. Vajdi and Pereira[84] did not observe any significant effect of radiation decontamination on the color and pigment content of paprika, while Byun and co-workers[133] noted some decrease in the acetone extractable pigment content and slight changes in Hunter's color values even at 5 kGy dose; a color difference, however, could not be noted by the naked eye between the untreated samples and samples irradiated up to 20-kGy dose. The latter authors found a very slight increase in the reducing and the total sugar content and a slight decrease of pH in red pepper by increasing doses in the 5- to 20-kGy range studied. An apparent increase in the hot water-soluble compounds was observed in pepper with increasing doses by Josimovic.[88] A dose of 15 kGy did not cause changes in paprika, black pepper, or a mixed seasoning detectable by dynamic thermal analysis.[134,135]

Galetto and co-workers[136] have studied the effect of irradiation treatment on chemical and nutritional constituents of onion powder. Attributes tested were volatiles, amino acids, hot water insolubles, color, and gross nutrients. At a minimum dose of 9 kGy, hot water insolubles were slightly reduced, while the optical index increased, although to a lesser extent than is caused by toasting. No gross nutritional changes were detected. The authors concluded that on the whole, onion powder appeared to be very resistant to chemical changes when treated with gamma irradiation. Gamma irradiation did not significantly change the nonvolatile oil content of garlic powder.[84]

The apparent increase of the volatile oil yield and the lipid content in some cases, as shown in Table 11, may be due to the large variation inherent in volatile oil estimation. Nevertheless, such observations as well as the increase of hot water solubles in pepper and onion powder mentioned earlier may also be related to a somewhat increased extractability of the plant tissue and due to the degradation of polysaccharides, i.e., pectin and starch (see also this chapter, Section III and Chapter 5, Section I), which are relatively easily attacked by irradiation. Starch is known to occur in some spices in high concentration (e.g., 20 to 40% in pepper) while pectin degradation is also a common cause of softening and increased juice-yield of fresh fruits and vegetables with higher radiation doses.[137] It remains to be seen whether this increased extractability of irradiated tissues, noted also on some medical herbs, teas, and other dry foods of plant origin, could be utilized technologically.

Production of H_2 and CO_2 was observed as an effect of gamma irradiation of various spices by Tjaberg et al.[106] The yield of H_2 with doses up to 45 kGy was proportional to the dose. G-values (the amount of gas produced per 100 eV absorbed radiation energy) varied between 0.5 and 1.7.

Measurement of weight loss under heating and analysis of aqueous extracts of seven spices (allspice, cinnamon, caraway, coriander, thyme, sweet paprika, and black pepper) for sugars, total carbonyls, and formaldehyde brought forward practically no changes after irradiation at 10 kGy.[88] No qualitative changes in absorption spectra of aqueous extracts of irradiated spices were observed.

Irradiation of dried parsley leaves with doses up to 50 kGy did not bring about any distinct qualitative or quantitative chemical changes.[138] No reduction of the chlorophyll content was noted up to 10 kGy. At 25- and 50-kGy dose levels, 3 and 6% decrease of chlorophyll appeared, respectively.

Summing up, dry products such as spices are less affected chemically by irradiation than items of high moisture content, and the aforementioned small changes occurring in some cases in the chemical composition of spices at dose levels sufficient for radiation decontamination seem to have no practical significance or speak in favor of the radiation treatment.

Table 13
EFFECT OF IRRADIATION OR FUMIGATION ON THE YIELD OF VOLATILE OILS FROM SPICES

	Yield of volatile oils as % of that of untreated spice														
	After radiation treatment (kGy)													After commercial EtO treatment	
Spice/condiment	2	4	5	7.5	8	10	12.5	15	16	20	30	45	50		Ref.
Allspice/pimento						97									83
			89					100				75			123
						100								26	84
Anise seed					100										139
Basil						99									83
Caraway seeds						88									124
			96	91		90	87	84		82	80		75		125
					111										139
Cardamom seeds			100	85		81	81	81							125
Celery seed						97									83
						100								100	84
Cinnamon						97								86	307
Cloves					98										139
Coriander			105					92				97			123
		120			120				120						19
			98												126
Cumin seed			118					116				126			123
						100-105									97
Fennel			117	92		108		83							99
			100	113		111		106							99
			92	109		123		105							99
			97												126
						98								105	307
Ginger						70									101
			87			66								100	307
					88										139
Juniper			84			71									73
					82										139

Mace					95							139
Marjoram			97			103						127
	109			126					169			73
			102				119			126		123
		100			100			100				19
Nutmegs			120			87						128
			111			90						128
						99					102	127
			99									104
			113									104
			102				108			91		123
			96									126
					100							139
Oregano						99						83
						100						83
			100								97	84
Papua nutmeg			100									126
Pepper, black						75						126
						101						83
			82			67						73
					92							139
			95				92			100		123
			90-100			95-112						117
						100					45	84
						98-100						97
		104			100			100				19
		100			100						70	320
Pepper, white						70						101
			104			102						127
			99									
			96									104
			100			100						96
					93							139
Thyme						101						83

Table 14
EFFECT OF IRRADIATION ON THE PERCENTAGE COMPOSITION OF ESSENTIAL OIL FROM CARDAMOM

		Dose (kGy)					
Peak	**Component**	**0**	**5**	**7.5**	**10**	**12.5**	**15**
a	α-Pinene	1.33	1.22	1.04	1.28	1.2	1.21
b	β-Pinene + sabinene	3.56	2.7	2.36	2.86	2.81	2.71
c		0.4	0.36	0.35	0.32	0.32	0.39
d		0.64	0.52	0.72	0.75	0.54	0.68
e	1.8 cineole (eucalyptol)	37.8	38.02	35.35	34.57	36.03	38.04
f		0.34	0.32	0.34	0.43	0.23	0.35
g		0.26	0.38	0.39	0.6	0.4	0.23
h		0.17	0.13	0.18	0.29	0.23	0.21
i	Linalool	2.34	2.03	2.63	2.01	2.03	2.09
j		0.24	0.2	0.23	0.31	0.17	0.25
k		0.22	0.24	0.18	0.20	0.28	0.23
l		1.08	0.95	1.28	0.94	0.94	0.93
m	Terpinenol-4	1.27	1.01	1.13	1.11	1.15	1.03
n		0.08	0.24	0.12	0.16	0.07	0.08
o	α-Terpineol	1.86	2.59	2.6	2.83	2.73	2.34
p		0.18	0.16	0.29	0.18	0.15	0.33
r		0.12	0.12	0.12	0.11	0.04	0.2
s	Terpinyl-acetate	46.77	47.31	49.1	49.79	49.45	47.34
t		0.33	0.45	0.22	0.13	0.20	0.22
u		0.31	0.34	0.36	0.34	0.32	0.34
w		0.07	0.1	0.35	0.09	0.07	0.11
z		0.54	0.57	0.61	0.63	0.56	0.61

From Bachman, S., Witkowski, S., and Žegota, A., in *Food Preservation by Irradiation*, Vol. 1, International Atomic Energy Agency, Vienna, 1978, 452. With permission.

Table 15
PERCENTAGE PROPORTION OF IDENTIFIED CONSTITUENTS OF VOLATILE OIL FROM IRRADIATED AND UNIRRADIATED SAMPLES OF WHITE PEPPER

		% Proportion in volatile oil	
Peak no.	**Identified as**	**0 kGy**	**10 kGy**
1	α-Pinene	5.3	5.3
2	Camphen	<0.1	<0.1
3	β-Pinene	9.9	9.8
5	4 (+) -3-carene	24.9	24.9
6	Phellandrene	4.8	4.6
7	Limonene	16.8	16.9
10	Terpinene	0.9	1.0
13	Linalool	<0.1	<0.1
15	Caryophyllene	27.8	28.0
16	Citral	1.2	1.2

From Weber, H., *Fleischwirtschaft*, 63(6), 5, 1983. With permission.

Table 16
COMPOSITION OF CINNAMON BARK OIL IN BOTH CONTROL AND IRRADIATED (10 kGy) SAMPLES

Peak no.	Constituent	% Control	1 Mrad
1	—	0.0226	0.0628
3	—	0.0621	0.1359
8	Limonese + 1,8 cineole	0.2656	0.0699
10	β-Cymene	0.2191	0.0173
15	Linalool	1.8908	0.1363
16	—	0.4222	0.0630
17	β-Caryophyllene	0.9674	1.0780
22	α-Terpenol	1.1308	0.1565
25	Cuminaldehyde	0.3033	0.0644
30	—	0.0439	0.3156
35	Cinnamaldehyde	61.1918	28.3930
36	Methyl cinnamate	1.3014	—
37	Cinnamyl acetate + eugenol	17.596	44.9403
38	Eugenyl acetate	3.2090	1.5146
39	(X) unidentified	1.899	
40	Cinnamyl alcohol	2.6084	18.7697
45	Benzyl benzoate	2.2586	0.7134

From Theivendirarajah, K. and Jayewardene, A. L., The Effect of Gamma Irradiation (^{60}Co) on Spices and Red Onion, IAEA Res. Contract No. 2840/JN, Food and Agric. Organ./Int. At. Energy Agency Res. Coordination Meet. on the Asian Regional Cooperative Project in Food Irradiation, Bangkok, November 22 to 26, 1982. With permission.

Table 17
EFFECTS OF IRRADIATION ON THE YIELD OF LIPIDS AND FATTY ACIDS IN CARDAMOM

			Dose (kGy)							
Peak	Component		0	5	7.5	10	12.5	15	30	50
	The yield of lipids (dry weight %)		7.0	6.5	7.5	8.0	7.5	8.0	8.5	8.8
A	Palmitic acid	$C_{16}^{:0}$	29.0	28.0	28.5	29.0	28.0	28.4	28.6	28.9
B		$C_{16}^{:1}$	2.5	3.4	3.7	3.8	2.9	4.1	3.5	3.0
C	Stearic acid	$C_{18}^{:0}$	2.9	4.0	4.1	3.5	3.2	4.0	3.3	3.7
D	Oleic acid	$C_{18}^{:1}$	44.5	44.1	44.0	45.0	44.2	44.3	44.0	44.5
E	Linoleic acid	$C_{18}^{:2}$	15.8	15.5	15.0	14.3	15.2	15.7	15.3	15.7
F	Linolenic acid	$C_{18}^{:3}$	5.3	5.0	4.7	5.4	5.5	5.5	5.3	4.2

From Bachman, S., Witkowski, S., and Zegota, A., in *Food Preservation by Irradiation,* Vol. 1, International Atomic Energy Agency, Vienna, 1978, 453. With permission.

Table 18
EFFECTS OF IRRADIATION ON THE YIELD OF LIPIDS AND CHARACTERISTIC FAT NUMBERS OF CARAWAY SEEDS

Dose (kGy)	0	5	7.5	10	12.5	15	20	30	50
Lipids (% dry weight)	17.0	17.4	17.9	18.2	18.3	18.4	18.6	18.6	17.5
Fat no.									
Acid no.	4.0	4.0	4.06	4.10	4.16	4.35	4.20	4.21	4.3
Iodine	11.7	11.6	11.5	11.5	11.4	11.3	11.5	11.2	11.0
Peroxide number	8.04	8.06	8.08	8.10	8.10	8.15	8.30	8.45	8.7

From Bachman, S., Witkowski, S., and Zegota, A., in *Food Preservation by Irradiation,* Vol. 1, International Atomic Energy Agency, Vienna, 1978, 446. With permission.

E. Effect of Irradiation on Organoleptic Qualities of Spices

While the objective of irradiating spices is reduction of microbial levels it is of prime importance that their flavor should not be adversely affected by the process. ''Sterilizing doses'' of 15 to 20 kGy may slightly or noticeably change the flavor characteristics of some spices.[87,119,140] However, the doses of 3 to 10 kGy, sufficient for ''pasteurization'', do not influence the sensory properties of an overwhelming majority of spices and herbs. Information on threshold doses causing organoleptic changes in spices and herbs is summarized in Table 19. Recent comparative flavor profile studies with highly diluted samples of selected spices (black pepper, paprika, onion powder, and garlic powder) showed less change in the flavor profile of radiation decontaminated than in fumigated samples.[320] It should be noted that even in cases where the sensitive methods of sensory panels may detect a statistically significant difference in flavor between untreated and radiation decontaminated spices, the spicing power is usually not influenced or not to an extent influencing the applicability of radiation decontaminated spices in the food industry. Various meat products prepared with spices that had been given a dose of up to 20 kGy could not be told by flavor from products prepared with the corresponding nonirradiated spices.[74,113] No significant differences were found in sensory properties between sausages made with irradiated or with EtO treated spices.[144] On the other hand, luncheon meat prepared with a number of different brands of spice extracts or with chemically decontaminated spices showed detectable off-flavors or flavors deviating from those of luncheon meat containing natural spices.[143] Experiments demonstrated that packaging and storage have a stronger effect on the sensory quality of spices than radiation treatment.[115,118,128]

Recently, scientists of McCormick and Co., Inc., (Hunt Valley, Md.) have done extensive sensory evaluation analyses for their entire line of spice products.[83] Various standard sensory panels were asked to test differences in sensory characteristics of irradiated and EtO-treated samples of allspice, Greek oregano, Spanish thyme, Mexican oregano, domestic paprika, Spanish paprika, celery seed, crushed red pepper, black pepper, garlic powder, and Egyptian basil. All irradiated samples were treated at an average dose of 10 kGy except for paprika and crushed red peppers, which were irradiated at an average dose of 6.5 kGy. No significant differences were found in any of the spices tested, except one sample of Spanish paprika. As a further check of the sensory properties of irradiated spices, all samples were evaluated in three types of food applications: hot, hot/cold (heated then refrigerated for 24 hr), and cold. ''In the applications tested, the irradiated spices were evaluated as having no differences in aroma and colour attributes when compared to the control. Slight differences in some flavours were noted. No major differences were found.''[83] Another company, Flavortech, reportedly found ''no significant alteration in the flavour profile of any of the spices tested'' at 26 kGy. There appeared to be a slight darkening in dehydrated onion and garlic products, but ''this factor was not considered critical in any of the finished products''.

Table 19
THRESHOLD DOSES CAUSING ORGANOLEPTIC CHANGES IN SPICES AND HERBS

Product	Threshold dose (T.D.) in kGy	Sensory testing	Ref.
Allspice (pimento)	~15.0	Triangular	73
Basil	>10.0		83
Caraway	~12.5	Triangular	73
Cardamom	~7.5	Triangular	73
Cayenne pepper	~10		141
Celery seed	>10.0		83
Charlock	~10.0	Triangular	73
Chive	4.0 < T.D. < 8.0		93
Cinnamon	>10.0	Paired comparisons triangular and rank test	94, 96, 307, 310
	<20.0		87
Cloves	<20.0		87
Coriander	<5.0 (odor)		96
	<5.0 but also >10.0	Triangular and rank test	94
	~7.5	Triangular	73
	>16.0		96
Cumin	6.0—10.0		97
Curry	>10.0		96
Dill tips	>10.0		98
Fennel	>10.0	Paired comparisons	307
	>15.0	Triangular	99
Fenugreek	<5.0 (odor)	Rank test in nonfat quarq	94
	>10.0	Paired comparisons	309
Garlic powder	3.0 < T.D. < 4.5 (aged sample)		141
	<10 (smell)		96
	>8.0 (odor, taste)		320
	>10.0 (fresh sample)		141
	>16.0		96
Ginger (dry)	>5.0		101
	>10.0	Paired comparisons	307, 310
	>45.0 (odor and flavor)	Dilution test	106
Juniper	>15.0		73
Lemon peel (powdered)	5.0 < T.D. < 10.0		94
Marjoram	<5.0 (odor)		96
	5.0 < T.D. < 10.0 but also >10.0	Triangular and rank test	94
	7.5 < T.D. < 12.5	Triangular	73
	>10.0		98
	>16.0		96
Mustard seed	>7.5		142
	>10.0		98
Nutmeg	>10.0	Triangular and rank test	89, 94, 128, 142
	>45.0 (odor and flavor)	Dilution test	106
Onion powder	<10.0 (optical index, color)		99, 110, 136
	8.0 < T.D. < 16.0 (smell)		110

Table 19 (continued)
THRESHOLD DOSES CAUSING ORGANOLEPTIC CHANGES IN SPICES AND HERBS

Product	Threshold dose (T.D.) in kGy	Sensory testing	Ref.
Onion powder	>8.0 (smell, taste)		320
	>10.0 (smell, taste)		98
	<16.0 (flavor)		96
Orange peel (powdered)	5.0 < T.D. < 10.0		94
Oregano	>10.0		83
Paprika	8.0 < T.D. < 10.0		89
	>8.0		320
	>10.0		96, 130
	>15.0		114
	>16.0		96
Pepper, black	<10.0		129
	>7.5		142
	>8.0		320
	>9.0		131
	~10.0		117
	~12.5	Triangular	73
	>10.0		96, 97, 119, 141
	>16.0		19
	<20.0		87
Pepper, white	>5.0		142
	>9.0		131
	>10.0		96
	>45.0		106
Red pepper	~10.0		141
	>10.0		83
Sage	>10.0		119
Thyme	≥10.0		94
	>10.0		83
Turmeric	5.0 < T.D. < 10.0	Triangular	94
	≥10.0	rank test	94

Varied results concerning the threshold dose of organoleptic changes as shown in Table 19 may be due to various reasons, including experimental errors. Radiation sensitivity may depend on the age of the seasoning being irradiated. Aged samples, where aroma qualities deteriorated in storage or typical aroma had been at a low level prior to decontamination treatment, may be more sensitive to radiation treatment than fresh, aroma-intensive batches. This was observed with garlic powder by Funke and co-workers[141] who found no difference between irradiated (up to 10 kGy) and control samples when fresh spice was irradiated; however, a significant difference was noted in the typical garlic-like aroma at doses of 4.5 kGy and upwards, if samples stored for 6 months at ambient temperature and 65% relative humidity were irradiated. Results of the same authors indicate that slight flavor changes may be masked by certain carrier media used in taste testing of spices.

Fenugreek proved to be one of the organoleptically radiation-sensitive spices, with a threshold dose of less than 5 kGy for sensory changes.[94] After irradiation at 5 or 10 kGy this spice lost its characteristic "herb-bitter" odor and showed a non-typical hay-like odor. The disadvantageous effect of radiation treatment could also be observed by taste testing in a carrier medium (curd from defatted milk containing 1% NaCl and 0.5% spice). Similar adverse organoleptic effects were noted by Zehnder and Ettel[94] on powders of orange and lemon peels where a hay-like, tea-like off-flavor appeared instead of their typical citrus flavor at 10-kGy dose.

Farkas and El-Nawavy[110] noted in experiments carried out with onion powder and onion flakes of Egyptian origin that with increasing radiation doses the characteristic smell of dried onions diminished and a caramel-like off-odor appeared and the color of the dried products darkened. These changes become significant only with treatment at 8 kGy (or higher) dose level. Differences between radiation-treated and -untreated samples, both packed in cellophane/polyethylene pouches, were eliminated (odor) or diminished (color) during 6 weeks of storage following irradiation. Results obtained by the COD method, based on bichromate oxidation of water-soluble reducing substances, were not in accord with the organoleptically observable changes of odor. In the U.S. commercial samples of onion powder were radiation decontaminated with 9 kGy causing sensory changes of practically negligible degree.[145] In Korean studies color difference between untreated onion and garlic powders and that of irradiated ones treated with 5 and 7 kGy could not be told by the naked eye, but it was distinguishable using Hunter color value.[146,305]

On the basis of the above survey one can conclude that radiation decontamination may induce some slight alterations in organoleptic properties of a few spices and condiments, but those changes are so slight that they barely are of practical importance in most of the cases.

F. Effect of Irradiation on Antimicrobial and Antioxidant Activities of Spices

Research results indicate that some spices have antimicrobial (mold inhibitory or bacteriostatic) properties and many of them act under specific conditions as antioxidants.[10] The major components causing inhibition of *Aspergillus parasiticus* were found to be cinnamaldehyde and *o*-methoxy-cinnamaldehyde in cinnamon and eugenol in clove.[334,335]

In Indian experiments radiation sterilization did not affect fungal inhibitory principles in clove, though marginal reduction was observed in that of cinnamon.[95] This may be due to a decrease in the cinnamaldehyde content being reduced from 61.8 to 45.5% in the steam-distilled oil from irradiated (10 kGy) sample as compared to the oil obtained from the unirradiated control. Eugenol levels were not affected by this radiation exposure.

In Japanese studies no changes in the antimicrobial activities of mace, sage, rosemary, clove, thyme, and oregano were observed up to 40-kGy irradiation.[18]

Antioxidant activities of various spices remained unaltered by radiation decontamination treatment.[18,147]

G. Herbal Teas and Dried Medicinal Plants

In some countries drug plants have become a significant market commodity. A large quantity of medicinal plant material is used in the preparation of herbal and medicinal infusions; three of the most widely used species for this purpose are two kinds of chamomile (*Matricaria chamomilla*, German chamomile, and *Anthemis nobilis,* Roman chamomile) and *Hibiscus sabdariffa*. It is estimated that at least 4000 ton/annum of chamomile imported from Argentina, Egypt, and Eastern Europe and 3500 ton of *H. sabdariffa* imported from the Sudan, Egypt, and Thailand are utilized for such infusions.[148]

Products such as ginseng, valerian, and licorice roots and a large range of herbs are a part of the fast-growing herbal and health food market. Medicinal plants are also becoming an important addition to certain industrial food products, especially dietetic preparations. Plants chiefly used are those which contain aromatic substances, vitamins, and important amino acids and enzymes to aid digestion and promote certain body functions.

Recent surveys of a number of dry tea herbs in various countries have also revealed a significant level of microbial contamination.[99,149-151] Some typical figures are given in Table

20. In a Yugoslavian study, total count as high as 10^7 to 10^8 microorganisms per gram was found in about 10% of chamomile samples investigated while about 80% of the samples contained between 10^5 and 10^8 microorganisms per gram.[149] *Salmonella* contamination of some tea herbs has occurred occasionally.[318,321] High incidence of *Salmonella* was reported in rooibos tea in South Africa.[328] Concerning fungal contamination, the main components of the mycoflora of herbal drugs seem to be those belonging to the genera of *Aspergillus* and *Penicillium*.[339]

Low microbiological contamination of herbs and medicinal plants is a necessity because no boiling, only maceration or infusion, can be applied to their preparation before consumption, because of their volatile or heat-sensitive essential oil ingredients. Thus, infusions might still contain relatively large numbers of microorganisms, jeopardizing both the safety of the consumer and the stability of the infusion.[150,153]

National Formulary regulations recommend the preservation of botanical crudes from insect infestation or microbiological contamination by means of suitable agents or processes that do not leave harmful residues. Radiation treatment may well become one of such processes.

Table 21 illustrates that in recent studies a radiation dose of 7.5 to 10 kGy was sufficient to reduce the total microbial count of several herb samples to the 10^3/g level or less. In a few cases a dose requirement of up to 15 kGy was noted when herbs were irradiated with accelerated electrons.[99]

The same doses effectively eliminate Enterobacteriaceae and molds, as shown by Tables 22 and 23.

The above doses necessary for radiation decontamination did not reduce the volatile oil content and did not significantly change the composition of volatile oils in herbs or the contents of coumarins and flavonoids in dry chamomile flowers.[99,126,150,153,154] The relative yields of volatile oils are given in Table 24.

Similar to some spices (see this chapter, Section II.D), a slight increase of the volatile oil yield can be noted in some cases. Although some of this may be due to the inherent variability of the estimation of volatile oil yield, perhaps it may also be explained by an increased extractability of the irradiated plant tissues.

No changes in the sensory properties of the infusions were observed up to 15 kGy.[99,150]

The feasibility of radiation decontamination of ginseng powder and concentrated chamomile extract was also reported by Korean and Yugoslavian authors, respectively.[154-156]

III. OTHER COMPLEX INGREDIENTS

A. Dehydrated Vegetables

Dehydration of vegetables represents a relatively small but important part of vegetable preservation. The majority of dried vegetables is utilized in the production of dry soups.

Because of great variation in the initial contamination of raw materials and differences in their processing technologies, the viable cell counts of dehydrated vegetables may vary considerably even under good hygienic conditions. Microbiologically, blanching is the most important cell count-reducing step in preparation of dehydrated vegetables. If it is properly done, it can practically eliminate heat-sensitive microorganisms, i.e., enterobacteria. Thermal inactivation during the dehydration process is less effective than usually expected.[157] Recontamination after blanching and/or dehydration should not be overlooked.[158]

Although dried vegetables are rarely involved in food-borne illness, spores of *Bacillus cereus, Clostridium botulinum,* and *C. perfringens,* if present in the soil, are likely to carry through into the final dried product and may become harmful if permitted to grow on reconstitution of such vegetables.[159] *B. cereus* is a frequent contaminant in dried vegetables, and its occurrence in dry potatoes or onion powder needs particular attention.[27,157] Heat

Table 20
MICROBIOLOGICAL CONTAMINATION OF SOME DRY HERBS (CFU/G)

Product	Total aerobic cell count	Enterobacteriaceae	Coliform	*E. coli*	*B. cereus*	Molds	Yeasts	Ref.
Anise	7×10^4		10^4			10^3—10^4	10^2—10^4	153
Balm	10^3—2×10^6		10^4			10^2—10^4	10^2—2×10^3	153
Chamomile	2×10^5	320		<20	300	160	160	99
	4×10^6	1×10^6		240	300	9000	2×10^4	99
	6×10^6	6×10^5		2200		1100		154
	10^3—2×10^7		10—> 10^4			10^2—2×10^4	10^2—6×10^3	153
Fennel	2×10^4	<20		<20	130	70	1200	99
	8×10^5		50			3×10^4	3×10^5	153
Hibiscus	10^4—2×10^5		<3			10^2—10^5	10^2—10^4	153
Hip	10^4—2×10^5		20—250			10^4	10^4	153
Lime blossom	2×10^5	3000		<20	350	590	<20	99
	10^5—10^7		10^2—10^4			10^3—6×10^4	10^3—7×10^4	153
Mint	4×10^6	2×10^6		<20	1×10^4	4×10^4	4×10^5	99
	10^5—3×10^7		10—10^4			10^2—2×10^6	10^2—3×10^6	153
Orange blossom	4×10^5		10^4			10^2	10^2	153
Orange leaves	6×10^5		10^4			1.5×10^5	5×10^3	153
Orange peels	5×10^3—3×10^4		20			10^2	10^2	153
Rooibos	4×10^7	2×10^7		<20	<20	1×10^5	2×10^5	99
Vervain	2×10^5	320		<20	300	260	160	99
	5×10^6		10^4			10^3—3×10^5	10^3—6×10^5	153

Table 21
EFFECTS OF IRRADIATION AND EtO FUMIGATION ON TOTAL AEROBIC VIABLE CELL COUNT (TAVC) OF SOME HERBS

	Log TAVC/g										
		Irradiated at(kGy)									
Product	**Untreated**	**3**	**5**	**6**	**7.5**	**8**	**9**	**10**	**15**	**Fumigated**	**Ref.**
Chamomile	7.9		4.6		3.5			2.3	<1.3		99
Flos chamomillae	7.1	6.0		4.0			3.1			4.4	150,153
	6.7						3.3			4.2	150,153
	6.6		5.6		5.3			5.0	3.0		99
	6.0							<1.0			150,153
	6.0		4.0			2.3		<1.3			150,153
	5.8	3.5	2.2								149
Fennel (Fruct. foeniculi)	4.3		2.6		2.5			1.9	<1.3		99
Hip	5.7	5.2		4.0			3.5			3.9	150,153
	4.6						3.3			3.4	150,153
Lime blossom	6.7					2.3					85
(Flos tillae)	6.4							<1.0			150,153
	6.0		3.0			2.4		1.6			150,153
	5.4	3.7		3.4			3.2			3.4	150,153
	5.3		3.1		2.2			1.7	1.7		99
	5.0									2.0	150,153
Mint	7.5						3.2			3.4	150,153
(Folia menthae pip)	7.1	6.0		4.8			3.2			4.7	150,153
	6.6		4.3		3.1			2.3	<1.3		99
	5.9		3.8			3.1		2.4			150,153
Rooibos (Asphaltus contaminatus)	7.6		6.3		4.3			4.0	2.5		99
Vervain	6.4					2.8					85
(Herb verbenae)	6.3						2.6			2.6	150,153
	6.2	5.0		4.8			3.1			2.5	150,153
	5.7		2.1			2.3		1.3			150,153
	5.3		3.0		2.4			1.3	1.3		99
	5.2							<1.0			150,153

sensitive nonsporulating microorganisms, i.e., *Escherichia coli* or *Salmonella,* should be destroyed by proper blanching, but those dried vegetables which were not previously blanched could contain these organisms if they were present in soils previously contaminated with animal or human wastes. Notably high incidence of *Salmonella* contamination has been found recently among samples of various dried vegetables (parsley — *S. livingstone;* carrots — *S. typhimurium;* mushrooms — *S. nashua, S. weltevreden, S. brunei, S. agona* and *S. derby;* and asparagus — *S. weltevrenden, S. muenchen, S. lanka, S. wirchow, S. agona,* and *S. anatum*) by Bockemühl and Wohlers.[318] Therefore, radiation decontamination of dehydrated vegetables may offer advantages to dried food processors or other dehydrated convenience food manufacturers.

The effects of irradiation on the total aerobic viable cell count of some dehydrated vegetables are shown in Table 25. It can be seen that a dose range of 4 to 8 kGy seems to be sufficient to reduce the aerobic plate count to the 10^3 to 10^4/g level, usually without changing the flavor qualities. In asparagus powder, some darkening and ''burned taste and flavour'' were noted at doses of 8 and 10 kGy.[98,99] Irradiation of mushroom powder or dry mushroom slices with 10 kGy of 10-MeV electrons did not influence the odor or taste, but resulted in some darkening.[99]

Table 22
EFFECTS OF IRRADIATION AND EtO FUMIGATION ON MOLD COUNTS OF SOME HERBS

Product	Untreated	Log mold counts/g Irradiated at (kGy)								Fumigated	Ref.
		3	5	6	7.5	8	9	10	15		
Chamomile	4.7							<1.0			150, 153
	4.0		2.9		3.2			2.3	<1.3		99
	3.9						2.2			3.4	150, 153
	3.5		<2.0								149
	3.3		<1.3			<1.3		<1.3			150, 153
	3.1	2.3		<1.0			2.0			<1.0	150, 153
	2.1		<1.3		1.3			<1.3	<1.3		99
Fennel	1.8		<1.3		<1.3			<1.3	<1.3		99
Hip	4.0						2.4			2.6	150, 153
	3.4	2.9		<1.0			2.7			2.6	150, 153
Lime blossom	4.3									2.8	150, 153
	4.3		2.1			1.3		1.3			150, 153
	3.5	3.1		2.6			<1.0			2.0	150, 153
	2.8		<1.3		<1.3			<1.3	<1.3		99
Mint	4.6		2.7		<1.3			<1.3	<1.3		99
	4.6						2.7			2.5	150, 153
	4.6		2.8		<1.3			<1.3			150, 153
	3.0	3.3		3.0			3.0			3.3	150, 153
Rooibos	5.0		4.3		3.2			3.3	1.3		99
Vervain	6.5							<1.0			150, 153
	4.3						3.0			<1.3	150, 153
	3.7	2.3		<1.0			<1.0		<1.0		150, 153
	3.3		<1.3			1.5		<1.0			150, 153
	2.4		<1.3		<1.3			<1.3	<1.3		99

Table 23
EFFECT OF IRRADIATION ON ENTEROBACTERIACEAE COUNTS OF SOME HERBS

Product	Untreated	Log Enterobacteriaceae count/g Irradiated at (kGy)					Ref.
		5	7.5	8	10	15	
Chamomile	6.0	5.0	4.7		4.0	1.3	99
	4.8	3.5		<1.3—1.7	<1.3		153
	4.6	2.6	1.6		<1.3	<1.3	99
Lime blossom	5.7	1.9		<1.3	<1.3		153
	3.5	<1.3	<1.3		<1.3	<1.3	99
Mint	6.3	3.9	<1.3		<1.3	<1.3	99
	5.3	1.6		<1.3	<1.3		153
Rooibos	7.3	6.0	4.3		2.6	1.7	99
Vervain	4.3	<1.3		<1.3	<1.3		153
	2.5	<1.3		<1.3	<1.3	<1.3	96

Table 24
YIELDS OF VOLATILE OILS FROM SOME IRRADIATED HERBS: AS PERCENTAGE OF VOLATILE OIL CONTENT OF UNTREATED CONTROLS

Herbs	Relative yields (%) after radiation treatment at (kGy)					
	5	7.5	10	15	25	Ref.
Chamomile	75	75	25	100		99
	103	97	134	128		99
	100					149
Fennel	117	92	108	83		99
	101	113	111	106		99
Mint	112		119	106		99
	110		108	99		99
			98		99	153
					92	153
Sage			106		103	153
Thyme			100		100	153

Table 25
EFFECTS OF IRRADIATION ON TOTAL AEROBIC VIABLE CELL COUNT (TAVC) OF SOME DEHYDRATED VEGETABLES

Product	Log TAVC/gram							
		Irradiated at (kGy)						
	Untreated	4	5	7.5	8	10	15	Ref.
Asparagus powder	6.0		4.8	4.3		4.0	3.0	98
	5.9				4.1	4.0		93
	5.6	2.9						85
Asparagus tips	6.7		5.1	4.3		3.0	2.0	98, 99
Carrots	4.8	2.7						85
Celery roots	5.0		2.0					160
Mushroom powder	4.9		3.3	3.0		<2.0		98, 99
	4.7	3.1						85
Mushroom slices	5.8		3.3	<3.0		<2.0		98, 99
Tomato powder	5.8				3.0			85
Yellow boletus								
Cut	5.3		3.7	3.0		2.3	2.3	98, 99
Ground	7.0		4.3	3.3		2.0	2.0	98, 99

In addition to microbial decontamination, radiation treatment of specific dehydrated items at 5- to 30-kGy dose range may have a tendering action, resulting in a decreased cooking time requirement for such dry vegetables as potato, cabbage, burdock, carrot, red gram, onion flakes, peas, and beans.[101,137,161-163,278,314] With onion powder, a significant increase of water uptake capacity was noted already at 4-kGy dose level.[320] The decreased cooking time is a consequence of decreased hardness and/or increased water absorption capacity due to partial breakdown of certain structural polymers and mobilization of calcium in the vegetable tissue.[101,163,278,280-282] By adjusting the irradiation dose for each vegetable, a uniform tenderness for all treated vegetables can be secured within a given cooking and rehydration period without significant off-flavor occurrence. This effect could be of value in dehydrated soup mixes. Considerable reduction of the cooking time of pea flour soup could already be obtained by a low dose of 3 kGy.[164] Here, evident changes of taste occurred with a dose of 4 kGy or above.

B. Dried Fruits

The microbiology and the preservation technology of dried fruits have been influenced in recent decades by the consumers' desire of a juicier product of higher moisture content (dried dates, figs, and prunes). Control of spoilage in such products is traditionally affected by chemical means or hot filling in suitable containers.[165,166] Because of the acidity of most fruits, only fungi can cause spoilage of dried fruits, and then only at relatively high a_w. Very few microorganisms can grow on fruits whose a_w is 0.60 to 0.70. If dried prunes are contaminated by *Xeromyces bisporus,* a mold that is able to grow at a_w near 0.60, spoilage may occur unless the fruit is preserved.[167]

Irradiation could control insect infestation and microbial spoilage in dried fruits as well. Experiments of Brower and Tilton[336] indicated that a dose of 0.4 kGy was effective in preventing insect development and feeding damage in dried fruits, i.e., raisins, zante currants, prunes, and dried apricots. If packages of nuts and dried fruits and vegetables contain only eggs and/or young larvae of beetles and moths, a dose of 0.2 kGy is adequate for control. Khan et al.[168] also investigated the effect of gamma radiation for the control of insects and microbial spoilage in some dried fruits, i.e., apricots, figs, dates, raisins, walnuts, almonds, chalghoza, and groundnut.

Doses exceeding 4 kGy can modify the texture of dry fruits resulting in reduced toughness and improved rehydration.[9,281,283] It is noted here that doses up to 4 kGy were found to increase the drying rate of blanched prunes.[170] Saravacos and Macris[337] reported that gamma irradiation of dried figs with doses up to 5 kGy caused no significant changes in color or in total and reducing sugars. Organoleptic evaluation of irradiated dried banana did not show off-flavor with doses up to 20 kGy.[284]

C. Dry Soup and Gravy Mixes

Dry soup and gravy mixes have many ingredients in common. These products are made simply by mixing dry commodities, then packaging them in laminated sachets or other moisture-proof containers. The main ingredients are meats, poultry, seafoods, vegetables, flours, starches and thickeners, fats, sugars, milk, and eggs.[171] The microflora of dry mixes depends on the flora of the dry ingredients.[36]

Dry mixes generally contain a wide variety of organisms. Aerobic plate counts generally are between 10^3 and 10^5/g.[172,173] *Staphylococcus aureus* or, more frequently, spore-forming *Clostridium perfringens* or *Bacillus cereus* are often present in dry mixes.[27,174,175] Dry gravy base used in food service establishments is rarely treated at high temperatures adequate to destroy bacterial spores. Although rarely, *Salmonella* contamination may also occur in dry soups.[176-178]

The microbiological quality of the so-called instant soups, which need not be cooked before consumption, is of particular importance.[179] If the reconstituted product is held warm, particularly between 30° and 50°C, eventual pathogens may grow to levels that will cause illness.[180-182] Instant soups shall, therefore, be formulated with ingredients which have a low bacterial load.

In addition to strict hygiene in preparation, radiation decontamination of critical ingredients or the dry mixes in the final package can be a reliable method for improving microbiological safety of such products.

D. Cereal Products

Besides their main use in making doughs, flours are also a basic ingredient in a wide

Table 26
MICROBIOLOGICAL PROFILE OF SOME CEREAL PRODUCTS[102,183,184]

	Range of viable cell count/g		
Microflora	Flours, cornmeals, corn grits, semolina	"Soy protein"	Dry cereal mixes
Bacteria			
Aerobic plate count	10^2—10^6	10^2—10^5	10^2—10^6
Aerobic thermophilic spores	10—10^2		
Coliform group	0—10	10^2—10^3	0—10^4
Escherichia coli		0—10^2	0—10^3
Psychrotrophs		10^2—10^4	
Clostridium perfringens		0—10^2	
"Rope" spores	0—10^2		
Molds	10^2—10^4		10^2—10^5
Yeast and yeast-like fungi	10—10^2		10^2—10^5

variety of sauces, gravies, sausages, meat loaves, canned foods, and confections. "Regular" microbial contamination of flours and similar cereal products is shown in Table 26.

Malts have higher microbial content than flours, e.g., 10^6 or more bacteria and yeasts per gram.[185]

Thermophilic spore-forming organisms are a part of the microflora of flours, and this is of particular interest to the processors of canned food. Ingredients of canned foods should meet the requirement of very low levels of flat sour and putrefactive spore-formers and sulfide spoilage organisms.[102]

Salmonellae have been detected often in soy meal and cottonseed meal and sometimes in other oilseed meals; their incidence in wheat flour is rare.[184]

Irradiation can be employed to reduce microbial levels in the above dry ingredients. Poisson et al.[186] reported that the bacterial count of wheat flour was quite considerably reduced with a dose of only 115 kGy, with less than 50% of the mold propagules (*Cladosporium, Penicillium,* and *Pullularia*) surviving. Vitamin B_1 content of the flour was not materially affected by this dose of irradiation. It is known that doses greater than 0.5 to 1 kGy can induce changes of wheat flour which impair baking qualities or organoleptic properties of products baked from them.[187] Utilization of ingredient mixtures containing irradiated flour did not impair, however, the sensory quality of canned meat products manufactured with such ingredients (see this chapter, Section VIII).

Zehnder and Ettel[188] reported that soy meal used as an ingredient of nut pastes could be effectively decontaminated with a radiation dose of 5 kGy. Although the sensory panel found statistically significant differences between the flavor of untreated and that of irradiated soy meal samples, the flavor change tended to be favored by the panelists.

Barley malt can be effectively pasteurized with ionizing radiations.[285] Malts irradiated with 1.1 kGy of gamma rays resulting in 99.9% reduction of viable bacteria were used in grain alcohol fermentations, saccharifying enzyme activity was not impaired, and no lactobacilli developed within 144 hr. Higher dosages of radiation resulted in substantially complete sterilization of the malts, but also in some destruction of their amylase activities.

Those cereal products which are used as ingredients in some dairy products and cannot be thermally processed because it would be detrimental to their sensory quality are also considered for radiation decontamination. For example, muesli-like cereal products mainly consisting of cereal flakes which are further mixed with sugar and fruit to bring a characteristic flavor and texture to a dairy product, sold as "complete breakfast", appear to be good candidates for radiation decontamination in France.[189]

E. Dried Egg and Milk Powder

Dried egg products are used in many ways in the preparation of processed foods (baked goods, soup, dry mixes, institutional foods, egg nog, ice cream, chocolate confections, pet foods, mayonnaise, macaroni, and noodles). The predominant organisms in dried eggs are enterococci and aerobic spore-forming bacilli. Salmonellae are the principal microbal problem. It is impossible to render egg products free from salmonellae by improved hygiene alone, because poultry themselves are known to be one of the biggest reservoirs of salmonellae. The problem becomes more serious if salmonellae grow during fermentation of the liquid egg for glucose removal. Although the number of salmonellae may be reduced as much as 4 log during drying, fermented albumin or temperature abused whole egg can have a high initial number so that survival of some salmonellae is likely.[190]

Despite the rather efficient procedures developed to pasteurize liquid egg, salmonellae appear occasionally in the final packaged product. Salmonellae remain viable in egg powder for a long time under storage conditions required for the maintenance of quality; therefore, the production of a *Salmonella*-free product is of primary importance. There is only a small safety margin between the thermal process that provides an adequate kill of salmonellae and a heat dosage that could seriously damage the functional properties of the egg product. Radiation may be considered as an alternative to the hot room treatment, which requires holding times of 7 to 14 days at 49° to 55°C.[190,191]

The survival of *S. typhimurium* and *S. senftenberg* as a function of electron radiation dose has been studied in early American investigations in dehydrated egg products.[192,193] There was no significant difference between these serotypes in dried whole egg, egg yolk, or egg white. From their data, Thornley[194] estimated the following dose requirements for *Salmonella* inactivation in various egg products to obtain a 10^7 reduction of viable count.

Dried whole egg	3.7 kGy
Dried egg yolk	5.7 kGy
Dried egg white	5.85 kGy
Sugared egg white (dried)	8.4 kGy

Using unflavored sponge cakes to evaluate flavor changes due to irradiation (approximately 7 kGy) in whole egg solids and egg yolk solids, a significant, but minor, effect of irradiation was noted. When unflavored custards were used as the test material the flavor differences were highly significant.[192]

For both egg yolk solids and whole egg solids, the volumes of sponge cakes prepared from irradiated material were somewhat lower than the volumes of cakes prepared from unirradiated materials. An increased whipping time was required to reach maximum volume when irradiated egg solids were used as compared with unirradiated solids.

Flavor tests on angel cakes prepared with unirradiated and irradiated (approximately 7 kGy) egg white solids showed no significant difference.[193] In case of sugared egg white powder there was a flavor difference significant at 5% level. The volume and texture of angel cakes prepared from egg white solids irradiated at approximately 7 kGy did not differ from cakes prepared with unirradiated commercial egg white. Irradiation had very little effect on foam stability. The solubility, viscosity, and ovomucin content of sugared and unsugared dehydrated egg white did not change significantly on irradiation. The fluorescence of irradiated products increased slightly on irradiation.

Kahan[195] reported results of radicidation tests on samples from a batch of dried egg albumin

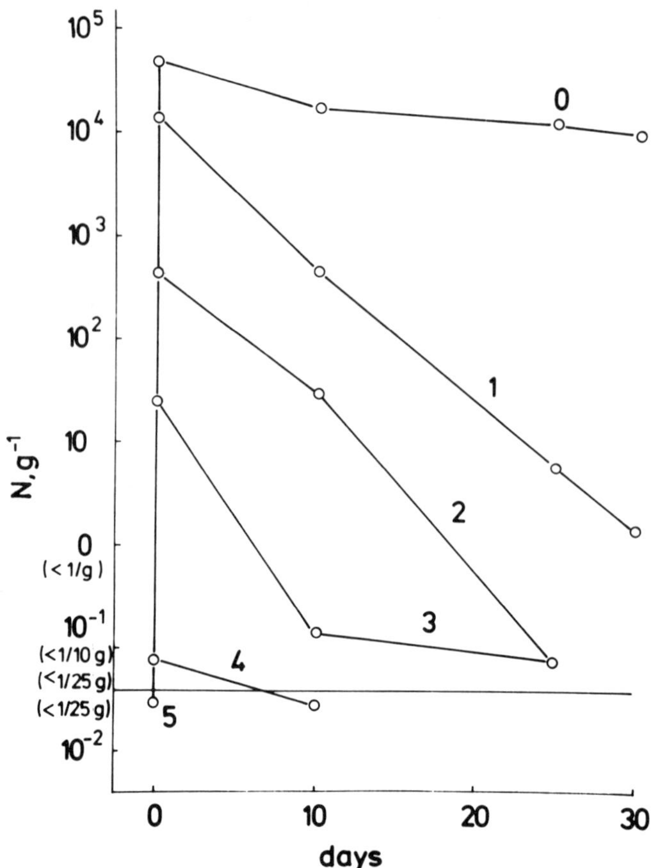

FIGURE 2. Influence of dose and storage time on the survival of *S. typhimurium, S. enteritidis,* and *S. lille* in whole egg powder at an initial contamination of 4.7 × 10^4/g. (From Katušin-Ražem, B., IAEA Res. Contract No. 3636/RB, Prog. Rep. December 1983 — August 1984, "Ruder Boskovič" Institute, Zagreb, Yugoslavia. With permission.)

that had been condemned for human consumption because of *Salmonella* contamination. Doses of 5 and 6 kGy were tested. *Salmonella* determinations were made about 3 weeks postirradiation. All the nonirradiated controls showed heavy *Salmonella* incidence, the 5-kGy treatment gave low *Salmonella* counts, and no *Salmonella* were detected in two of the four 6-kGy treatments. The author concluded that doses of about 6 kGy may be adequate for radicidation of dried egg albumin.

Italian authors observed exponential destruction of *S. typhimurium* by irradiation in dried egg albumin, with a D_{10} value of 1.07 kGy.[196]

In dried egg yolk artificially contaminated with *S. typhimurium*, radiation treatment with accelerated electrons reduced the viable count of *Salmonella* by approximately 99% at 2.5-kGy dose, and by more than 99.9% at 5-kGy dose level.[197] Destruction of *Salmonella* cells at an initial count of 2.7 × 10^5/g required a dose of 10 kGy when colony counts were estimated within 8 days after irradiation. However, after 100 days of storage of dried egg yolk, no viable *Salmonella* could be detected even at 1-kGy dose. This "after-effect" which has been demonstrated also in work on some laboratory animal feeds and spices at low moisture levels (see also this chapter, Section VIII.B) might be related to the microbetoxic effect of some long-lived radiation products (free radicals or their secunder products). Egg and milk powders were found to retain ESR signal for long time even at and above normal moisture levels.[198]

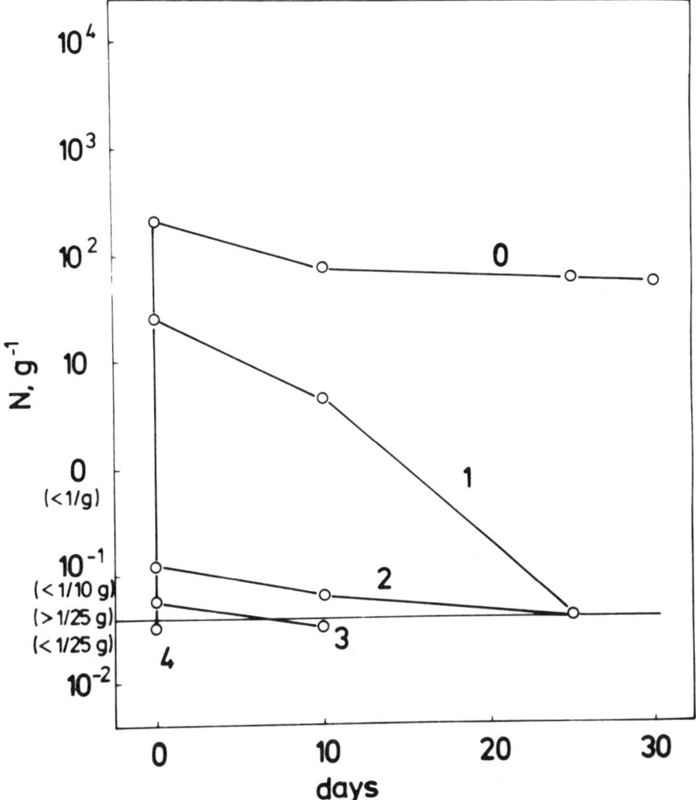

FIGURE 3. Influence of dose and storage time on the survival of *S. typhimurium, S. enteritidis,* and *S. lille* in whole egg powder at initial contamination of 3.7 × 10^2/g. (From Katušin-Ražem, B., IAEA Res. Contract No. 3636/RB, Prog. Rep. December 1983 — August 1984, "Ruder Boskovič" Institute, Zagreb, Yugoslavia. With permission.)

Similar after-effects have been found in recent Yugoslavian studies on *Salmonella* eradication on whole egg powder γ-irradiated under aerobic conditions.[199] Here, a mixture of lyophilized cultures of *S. typhimurium, S. enteritidis,* and *S. lille* was used for inoculation of commercial egg powder of 3.5% moisture content to an intended contamination of approximately 10^4 and 10^2/g, respectively. The results of microbiological investigations are given in Figures 2 and 3 showing the total number of three applied strains of *Salmonella* as a function of dose and storage time after irradiation at two applied levels of initial contamination.

The dose-survival curve of the mixed population is shown in Figure 4. From the linear part, a decimal reduction dose (D_{10}-value) of 0.8 kGy can be estimated.

Chemical analysis of fatty components showed that the percentage ratio of triglycerides in the lipid content decreased while the percentage of mono- and diglycerides increased with increasing doses. No change in the degree of acidity of lipid fraction extracted with diethyl ether was observed up to 10-kGy dose.

The influence of radiation treatment of the vitamin content and peroxide values directly after irradiation is given in Table 27. (The increase of peroxide value could be expected, due to peroxidation of unsaturated fatty acids during irradiation in the presence of air, as had already been observed in early studies.[192])

Chemiluminescence measurements also showed a higher degree of autooxidation at higher doses (5 and 10 kGy).

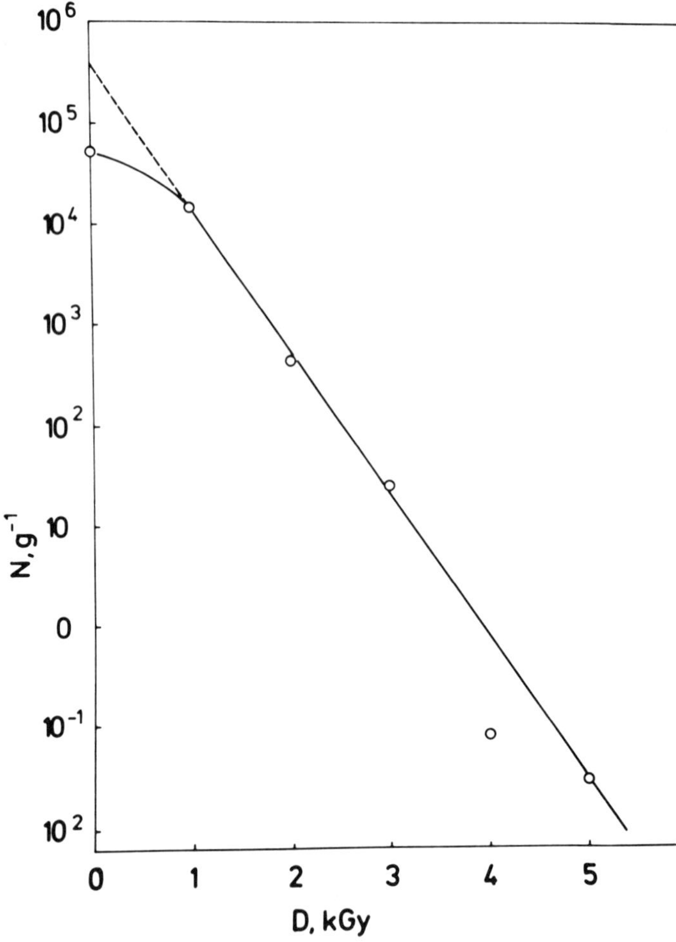

FIGURE 4. Dose-survival curve of the mixture of *S. typhimurium, S. enteritidis,* and *S. lille* in whole egg powder. (From Katušin-Ražem, B., IAEA Res. Contract No. 3636/RB, Prog. Rep. December 1983 — August 1984, "Ruder Boskovič" Institute, Zagreb, Yugoslavia. With permission.)

Table 27
EFFECT OF GAMMA RADIATION ON THE PEROXIDE VALUE AND VITAMIN CONTENTS IN WHOLE EGG POWDER UNDER AEROBIC IRRADIATION CONDITIONS[199]

Dose	Peroxide value	Vitamin A (I.U./100 g)	Thiamin (mg/100 g)	Riboflavin (mg/100 g)
0	0	805 ± 10	0.169	1.31 ± 0.04
1	0.75 ± 0.01	820		
2	1.50 ± 0.01	835		
3	2.22 ± 0.03	650		
4	3.23 ± 0.03	425 ± 25	0.169	1.33 ± 0.03
6	4.47 ± 0.04			
10	5.73 ± 0.05			

Note: Data are mean values of five determinations ± S.D.

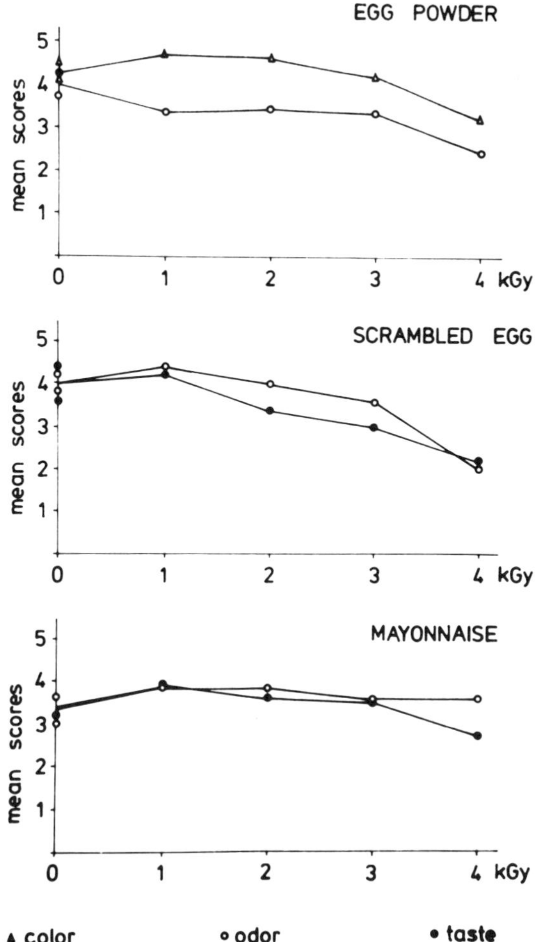

FIGURE 5. Sensory evaluation of irradiated whole egg powder and scrambled eggs and mayonnaise prepared from it. [199]

Radiation-induced loss of vitamin A in irradiated whole egg powder has already been investigated by Diehl[200] as a function of storage time after electron irradiation. Contrary to the Yugoslavian studies, he found that vitamin A was relatively stable against irradiation. This may be due to the large difference of dose rates (36 Gy/min for γ-irradiation and 10^8 Gy/sec for electron irradiation) in the respective studies. At very high dose rates peroxidation has much less chance to develop.

Radiation-induced degradation of several amino acids, particularly methionine, histidine, tyrosine, and lysine, found to be very reactive with hydrated electrons and hydroxyl radicals was also noted in the composition of the acid hydrolysates of egg powder samples at 5 and 10 kGY.[199]

Organoleptic evaluation of egg powder and odor and taste of scrambled egg and mayonnaise prepared from the samples has shown that samples irradiated up to 2 kGy could not be distinguished by their sensory properties from unirradiated ones. Sensory properties started to deteriorate at 3-kGy dose level, with clearly inferior quality at 4 kGy (Figure 5).[199]

From the literature survey summarized above it can be concluded that off-flavor seems to be the limiting factor in radiation decontamination of dried whole egg irradiated under aerobic conditions.

It remains to be seen how well anoxic conditions could prevent radiation-induced oxidative changes and off-flavor formation in irradiated whole egg and dried egg yolk. Dried egg yolk albumin, being less sensitive to autooxidation than egg yolk containing fatty products, seems to be an easier item for radiation.

The potential hazard of milk powder as a source of *Salmonella* infection is of concern. Numerous lots of milk powder have been suspected or confirmed as vehicles of infection in several outbreaks.[201] Unfortunately, whole milk powder is even more radiation sensitive than whole egg powder. Under aerobic irradiation already doses of 0.3 to 0.6 kGy were found to result in off-flavor and oxidation of milk fat with increased peroxide and TBA values and to reduce vitamin A and carotenoids content significantly.[202,316,317] Autooxidation of milk fat and losses of vitamin A and carotenes proceed faster in irradiated than in untreated samples.[202] Irradiation of skim milk powder at a relatively low dosage level also produced volatiles capable of inducing off-flavors in recombined skim milk.[286,287] However, the incorporation of radical scavengers, ascorbic acid, and ascorbyl palmytate may permit dosage treatments of a level known to eliminate *Salmonella.*

F. Cocoa Powder and Desiccated Coconut

The increased use of cocoa powder in chocolate, chocolate coatings, and in several chocolate food drinks has stimulated the market for imported cocoa powder.

Microbial populations of cocoa vary tremendously, possibly reflecting the microbial quality of the beans and the technology of the country in which the beans were roasted and processed.[203] A Dutch study of 547 samples of cocoa powder showed the *Bacillus subtilis* and *B. licheniformis* predominated as aerobic spore-formers.[204] In another study *B. cereus* accounted for approximately 20% of all isolated bacilli.[205]

Although *Salmonella* are expected to be killed by adequate roasting, they have been found in cocoa in several occasions, and they caused illness in a large number of consumers.[206,207] Yeast and mold counts in cocoa packaged for the retail trade may be high (up to 10^5/g), although most recommended limits are less than 50/g.[203]

Irradiation of cocoa powder, although it destroyed microbes, affected the flavor of the product adversely.[208] A dose of 5 kGy is only acceptable if irradiation is performed under vacuum or in an inert atmosphere. The sensitivity of cocoa powder to irradiation increased with fat content.

Inadequate sanitary controls during the harvesting and processing of coconut contribute to *Salmonella* contamination of desiccated coconut.[206] Decontamination of shredded coconut by roasting was ineffective and led to discoloration of the product. Kovacs[332] considered decontamination of coconut by irradiation in 1959. Although in later studies gamma radiation doses of 1.3 to 1.6 kGy were 90% effective in eliminating five *Salmonella* serotypes in artificially contaminated coconut, 4.5 kGy was required to disinfect coconut affecting organoleptic characteristics adversely.[210]

G. Coloring Substances

From the natural coloring agents, paprika, which is not only a spice but also a coloring substance, has been dealt with in this chapter, Section II.

Powdered, defatted shell of cocoa beans used as a coloring agent in the food and cosmetics industry can be successfully decontaminated by irradiation.[93]

Apparently, various synthetic pigments applied in the pharmaceutical industry can be radiation sterilized without adverse changes.[295] A series of selected pharmaceutical colorants (ten water soluble and nine insoluble) were subjected to a 25-kGy radiation dose and exhibited no changes when examined by visible, UV, and IR spectrophotometry. TLC with four solvent systems showed no difference between irradiated and unirradiated colorants.[326] Red D & C azo pigment (GBL-70-219) was shown by Prince and Welt[327] to undergo no visible

Table 28
INCIDENCE OF *BACILLUS* SPORES IN STARCH SAMPLES TESTED IN JAPAN[213]

After heating at	No. of samples tested	No. of spores/g		No. of negatives
		$\leq 10^1$	10^2	
75°C for 20 min	42	16	11	15
100°C for 10 min	42		2	40

change when irradiated at 50 kGy. Similarly, FD & C Red No. 3 powder at 15 kGy shows no change in UV absorbance.

IV. TEXTURIZING AGENTS AND PROTEIN PREPARATIONS

A. Vegetable Gums and Agar

Some plant hydrocolloids (vegetable gums) used as thickening agents seem to be more sensitive to radiation treatment than multicomponent spices or dry vegetables. Particularly guar and carob gums show impairment of rheological properties due to fragmentation into smaller molecules even at doses as low as 0.2 to 0.5 kGy.[98,99,188] The undesired reduction of viscosity may be compensated for, however, by a slight increase in the concentration, since the consistency of the aqueous solutions of the gums increases exponentially with the concentration.[98]

In gum arabic (gum acacia) a considerable reduction of microbial contamination can be achieved with relatively low doses which do not seriously harm the functional properties.[325] According to Henon,[189] a medium dose of irradiation is more effective than EtO to destroy molds and yeasts in gum arabic.

Radiation decontamination of tragacanth may also be achieved under relatively small changes of its viscosity.[211] It was reported that at 10 kGy, the decrease in the coefficient of viscosity was only 7%.[324]

It has been found that agar is less sensitive to irradiation than other thickening agents.[212]

B. Starches

Microbiological purity is one of the important quality characteristics of starch preparations intended for the food processing industry. Even though the viable counts are rather low, the contamination of starches with thermophilic bacterial spores proved to be quite bothersome, particularly to the canners.[102]

Table 28 shows the incidence of *Bacillus* spores in commercial samples of starch from a recent Japanese study.[212] The distribution of species of aerobic spore-formers in the same study is shown in Table 29. *B. licheniformis* spores were most encountered also in French studies.[216]

French workers were able to show that the viable cell counts of commercial starches could be satisfactorily reduced by doses around 3.0 to 4.0 kGy.[214-219] Delattre et al.[219] published a nomogram for estimation of the dose requirement of radiation treatment as a function of the initial contamination of starch and permissible levels of survivors. Over 4 kGy, considerable changes in viscosity, reducing capacity, pH, and iodine-binding capacity occur. In the 10-kGy range, dextrin formation and production of monosaccharides as well as the Maillard reaction may also take place.[220] At high doses, technologically interesting new properties could develop in irradiated starches since their molecular weight, adhesive capacity, and water solubility were altered and could be controlled.[212,221] Such starches inhibit

Table 29
DISTRIBUTION OF SPECIES OF *BACILLUS* SPORES OBTAINED FROM STARCH[213]

Species	No. of isolates
B. licheniformis	22
B. megaterium	18
B. subtilis	13
B. coagulans	4
B. polymyxa	3
B. laterosporus	3
B. macerans	3
B. cereus	2
B. sphaericus	1
B. alvei	1
B. brevis	1
Total	71

lead and steel corrosion in water and in malic acid solution. The possibility exists for using irradiation to obtain modified starches which suit specific applications in the food, textile, paper, and other industries.[288-290]

C. Pectin

In the dry state, radiation decontamination of pectin can be carried out without undue changes in the molecule. Even at a dose level of approximately 20 kGy, only small amounts of dialyzable, i.e., small-size, breakdown products are formed. In contrast to this, the specific viscosity, i.e., molecular size, is reduced by approximately 40%. Doses around 5 kGy applied in the dry state do not affect viscosity seriously.[222] Nevertheless, radiation treatment has been suggested also to produce pectins with modified gelling properties.[281,291,292]

D. Gelatin

The microbiological purity of gelatin is important for its various applications in meat products and in the pharmaceutical and photofilm industries.

The microbiological quality of dry gelatin can be effectively improved by 5 to 10 kGy without seriously affecting technological or organoleptic properties.[223-225] However, the threshold dose for a flavor change and decreasing viscosity seems to be approximately 5 kGy. With doses higher than 5 kGy a peptone odor is produced which changes into a broth odor (above 25 kGy) and then to the odor and taste of bone glue (above 50 kGy).[225] Under anoxic irradiation conditions the threshold dose for flavor changes is higher than 10 kGy.[224]

Sterilization of dry gelatin can be secured at a dose of 15 to 35 kGy.[224-226] In the early studies of Mateles and Goldblith[227] no reduction in the gel strength of irradiated gelatin was noted at doses up to approximately 17 kGy studied. Frank and Grünewald[226] reported a reduction of gel firmness by one third of the original value. Water solubility was not appreciably altered while the speed of gelling dropped somewhat. Other authors reported 15 to 35% reduction of the specific viscosity of gelatin after 20 kGy, due to a decrease in chain length, although the gel strength remained within British Pharmacopoeial limits. Irradiation of gelatin did not produce any distinct changes in amino acid composition up to 35 kGy though it did give rise to an increase in the content of carbonyl groups.[225] At higher doses (50 to 60 kGy) growth of some bacteria is inhibited in media containing irradiated gelatin.[226]

E. Protein Preparations

Mainly Argentinian, Bulgarian, and French studies demonstrated that protein preparations of both animal and plant origin (powdered whey, sodium caseinate, dehydrated blood and plasma, soybean protein, etc.) may be treated with 5- to 15-kGy doses which effectively decontaminate them without appreciably altering physicochemical and organoleptic characteristics.[228-230,237] Sterilization required a dose of 20 kGy or higher.

Defatted fish protein concentrate, which is used as the flavoring agent in pet foods and is considered a potential source of salmonellosis in animals, has a moderate total viable cell count ($\leq 5 \times 10^4$/g) of microorganisms. Therefore, it can be effectively decontaminated with a minimum dose of 5 kGy.[231] A survey of 59 lots of irradiated fish protein concentrate powder, received over an 18-month period, indicated that the microbial counts of all the lots were below 10 cells per gram.

V. SUGAR, SALT, AND TALC

A. Sugar

While its microbial contamination is usually low in numbers and it does not pose a health hazard, its microflora may impair the keeping quality of products of the utilizing industries. Yeast and molds can cause troubles in the beverage and confectionary industries and thermoresistant bacterial spores may deteriorate the quality of the product from the canners' point of view.[232] Anaerobic spore-forming bacteria *Clostridium thermosaccharolyticum* and *Desulfotomaculum nigrificans* as well as the so-called flat-sour bacteria *Bacillus stearothermophilus and B. coagulans* have to be killed in the sugar to obtain a product which can be used by the canning industry. If present insufficient numbers in the sugar they cause serious spoilage in foods that contain sugar as an ingredient.[233]

Commercial granulated beet sugar can be sterilized by radiation doses of 10 to 20 kGy.[234,235] Probably much lower doses are sufficient to eliminate the thermophilic spores present in low number. Canned peas prepared with a brine containing radappertized sugar did not show flat-sour spoilage in contrast to canned peas prepared with nonirradiated sucrose. At doses larger than 5 kGy, various degradation products can be detected, and a pinkish discoloration of the solid sugar can be observed. The UV spectrum of the solution of irradiated sugar showed a maximum at 265 nm. This was accompanied by increase in the reducing power especially in samples which had been heat treated in addition to radiation. The presence of considerable amounts of glucose and fructose could be detected by paper chromatography. The discoloration of irradiated sugar may be due to irradiation-induced caramelization.[98] Some of the reaction products have specific microbiological effects.[235] Radiation sterilization of lactose and dextrose preparations have been tried by the pharmaceutical industry.[293] Sterility could be achieved with doses up to 25 kGy, but it was accompanied with significant color changes.

B. Salt

Salt may also constitute a microbial contamination source when unpurified table salt is used in food. Frequently salt is a component of spice mixtures. Very little work seems to have been done on radiation decontamination of salt. Grünewald[98] studied the effect of electron irradiation of the color of food grade salt. He found pronounced changes in color due to irradiation. Irradiated salt samples had, in comparison to the nonirradiated sample, a brownish color. The intensity of discoloration increased with the dose. If the irradiated samples were exposed to daylight the brownish color changed to gray. These color changes of irradiated salt are based merely on lattice defects in the salt crystals and not on chemical changes of the salt, because the salt lost its brown or gray color immediately upon dissolution in water, the solutions of nonirradiated and irradiated salt had identical absorption curves in the range between 200 and 800 nm, and the recrystallized salt showed no difference in color as compared to the nonirradiated sample.

It was shown in academic studies, which help in understanding the mechanism of the above phenomena, that the dissolution of γ-irradiated sodium chloride in pure water and aqueous sodium nitrate solution results in the emission of light, in the former, and the formation of nitrite, in the latter case.[236] This aqualuminescence on the dissolution of the defect centers is due to release of electrons which form e^-_{aq}, and these attack NO_3^- ion to give rise to NO_2^-. Exposing irradiated sodium chloride to visible light is sufficient to bleach the defect centers (photoannealing).

C. Talc

For various purposes, talc is used both in pharmaceutical and food industries. Radiation sterilization of talc is commercially established in several countries for the pharmaceutical industry. An increase in the acid-soluble matter of talc has been reported at 25 kGy although no other change was noticed.[293] The color and British Pharmaceutical Codex test for alkalinity were unaffected.

VI. INDUSTRIAL ENZYME PREPARATIONS

Enzyme preparations are increasingly used both in the food and pharmaceutical industries. Many of these preparations are of microbial origin. Crude enzyme preparations usually contain microbial contamination which may become a possible danger even in the food processing industry for the products treated with the enzyme.[238,244] Therefore, it is required that not only clinically used but also industry-aimed enzyme preparations, apart from being free of pathogenic or toxic microorganisms, exhibit a low microbial contamination (maximum 10^2 to 10^4 saprophytic cells per gram).

Due to the high radiation resistance of enzymes in dry form or in their natural environment, the reduction of microbial count in industrial enzyme preparations by ionizing radiation is another viable application area. Solid enzyme preparations have been found to be more stable than enzymes in solutions and more radiation resistant than microorganisms. After the pioneering work of Vas on radiation disinfection of pectolytic preparations, it has been shown by several laboratories that radiation doses of around 10 kGy or less are sufficient to reduce microbial load of fungal polygalacturonase and lipase and bacterial proteases and amylases to negligible (10^2/g) levels without appreciably affecting enzyme activity.[239-249,296-298,329]

Structural changes in the dry enzymes after irradiation could not be detected by electrophoretic or chromatographic methods. Specificity of papain and a *Mucor pusillus* protease was not affected by irradiation.[300,301]

Many enzymes are quite radiation resistant. Dry pepsin, papain, ficin, and Maya protease could be microbiologically decontaminated by radiation doses of up to 20 to 30 kGy with practically unchanged enzymatic activity.[250,251,294,295] EtO treatment caused approximately 80% reduction both in the proteolytic and the estherase activity of dry papain.[252]

Russian authors studied the possible use of ionizing radiation combined with a variable magnetic field of 750-Oe intensity for 18 hr or with mild heating at 50°C for 1 hr to sterilize proteolytic enzymes (Amylorizin P10x, papain, trypsin, and chymotrypsin).[253] Gamma irradiation at 10 kGy combined with one of these treatments resulted in a lower bacterial survival rate than by γ-irradiation alone while completely maintaining the proteolytic activity of the enzymic preparations.

Others noted more or less decrease in the enzyme activity of various preparations as a function of the radiation dose.[254-256,296-298] As one can expect, purification increases the susceptibility of enzyme preparations to irradiation. Some of the microbiological and enzymological data reported are summarized in Table 30.

The relatively large differences in the radiation resistance of various enzyme preparations may be due to the grade of their purity, the presence or absence of some protecting substances, etc.[243,259]

The temperature has a considerable influence on the radiation sensitivity of both enzymes and microorganisms. Because low temperatures have less protective effect on enzymes than on microbes, in such cases when radiation decontamination at room temperature causes a notable reduction of enzyme activity, it is worthwhile to consider performing the radiation treatment under the freezing point.[299]

Increased heat sensitivity of enzymes was reported by several authors after irradiation of enzyme solutions.[244,301,302] One can assume, however, that in dry preparations this heat sensitizing effect is much less at pasteurizing doses.

Although more studies are of interest, e.g., on storage stability of irradiated enzyme preparations (see this chapter, Section VIII), the feasibility of using irradiation for decontamination of enzyme preparations with similar doses as those established for spices looks promising. In fact, some enzymes, e.g., proteases used in detergents have been irradiated routinely on a large scale for more than 25 years.[294]

Prevention of microbial growth in bioreactors utilizing immobilized enzymes or immobilized cells is of eminent importance. Bachman and Gebicka[257] irradiated glucose isomerase contained in *Actinoplanes missouriensis* and *Streptomyces olivaceus* cells, respectively, in the dry state. Approximately 5 and 10% loss of enzyme activity occurred at 10 kGy in the two kinds of preparations of whole cell glucose isomerase. Radiation treatment up to 10 kGy had no significant influence on intracell binding of both glucose isomerases and on their enzymatic properties (optimal temperature, thermal stability, optimal pH, pH stability, and requirement for metal ions).[258]

Although it is not directly related to the topic of radiation decontamination, it may be noted here that radiation treatment was successfully utilized to immobilize enzymes by radiopolymerization with monomers, such as acrilamide. Such entrapment is effectively increased by copolymerization with soluble starch and lyophylization treatment (or in the frozen state).[303] In this way various forms on immobilized preparations can be easily prepared which have a spongy texture with a high surface area.

VII. INCREASED SENSITIVITY OF THE RESIDUAL MICROFLORA SURVIVING RADIATION TREATMENT

Surviving microbial flora of ingredients treated with a ''pasteurizing'' dose of radiation is proved to be sensitized to further antimicrobial actions and certain environmental effects. The survivors have lower heat resistance and salt tolerance, and they are more demanding as regards their pH, moisture, and growth temperature requirements than the microorganisms of untreated ingredients.[74,114,260,262] Some of these effects are shown in Table 31 illustrating the heat and salt sensitivity in the microbial flora of a mixed seasoning surviving a radiation treatment of 3 kGy.[74] The mixed seasoning, either untreated or irradiated, was suspended in the ratio 1:30 in a liquid medium containing various salt concentrations and was subjected to heat treatment at 80°C for 10 min or plated without heat treatment. Numbers of colony forming units (N) were determined by plating dilutions into nutrient agar of identical composition and salt concentration to those of the suspending media. The pH was set at a value of 6.5.

In comparative investigation on the effects of gamma radiation and EtO fumigation on the bacterial spore flora of black pepper, an increased sensitivity to the reduction of the pH of the recovery medium from pH 6.0 to 5.0 was observed in the survivors of the radiation treatment, whereas the survivors of fumigation did not show significant differences in the colony counts in this pH range.[77,263]

Table 30
EFFECTS OF IRRADIATION ON THE VIABLE CELL COUNT AND ACTIVITY OF SOME ENZYME PREPARATIONS[a]

Enzyme preparation	Untreated	Irradiated at (kGy) 2.5	3	5	7.5	9	10	15	20	40	Ref.
α-amylase (bacterial)	4.9			4.0/97			1.3/93				249
α-amylase (fungal)	3.8		<2.0/99	<1.0/102	—/94		—/94				243
	3.8		<2.0/99	<1.0/102	—/94		—/94				
Pancreatin A	5.2	4.5		3.8			<1.0	—			295
Amylolytic act.		4.5/96		3.8/99			<1.0/94	—/87			
Lipolytic act.		4.5/99		3.8/99			<1.0/89	—/94			
Proteolytic act.		4.5/96		3.8/95			<1.0/93	—/96			
Pancreatin B	6.3			3.1			2.1	1.7			295
Amylolytic act.				3.1/98			2.1/92	1.7/93			
Lipolytic act.				3.1/93			2.1/73	1.7/73			
Proteolytic act.				3.1/97			2.1/93	1.7/83			
Pancreatin C	3.9			1.7			—	—			295
Amylolytic act.				1.7/98			—/96	—/94			
Lipolytic act.				1.7/98			—/94	—/87			
Proteolytic act.				1.7/91			—/89	—/84			
Pancreatin (techn.) lipase	4.3			2.1/95	—/85		—/76	—/66			256
Pectinase	5.3		3.9/93	2.7/91-105	<2.0/92-97		<1.0/90-91				243
	4.7		3.6/93	2.6/91-105			<1.0/90-91				
Endopolygalacturonase	7.7		3.3/94				1.6/77		1.5/70	<0.8/60	255
Polygalacturonase	8.0					4.0/82					242
	7.0					2.3/86					
	9.0					4.3/102					
Protease (bacterial)	5.6		3.6/99	2.3/96	<2.0/95		<1.0/93				243
	7.6		5.9/99	4.2/96			<1.0/93				
Protease (crude) from *B. subtilis*	10.7			6.0/92			4.8/120				279
	6.9	4.2/113			2.9/100						
	9.7						5.7/113		3.5/133		

Protease		5.3	<2.0/99-100	<1.0/95-101	—/98-102	—/91-100		243
(fungal)		5.1	<2.0	<1.0		—/91-100		
Protease from	I	9.5		4.7/100		—/98	—/91	245
Thermoactinomyces	II	3.6		—/100		—/98	—/91	
vulgaris	I	6.9		—/100		—/100	—/98	
	II	3.9		—/100		—/100	—/98	
	I	5.8				—/96		
	II	4.7				—/96		
	I	5.0				—/95		
	II	5.7				—/95		

Note: I, *T. vulgaris*; II, aerobic mesophyllic bacteria.

[a] Log CFU/g/% residual enzyme activity.

Table 31
HEAT AND SALT SENSITIVITY IN THE MICROBIAL FLORA OF MIXED SEASONING SURVIVING RADIATION TREATMENT

		Log number of cells/g capable of colony formation					
		Unirradiated			Irradiated at 3 kGy		
		0.5% NaCl	5% NaCl	10% NaCl	0.5% NaCl	5% NaCl	10% NaCl
		Water activity (a_w)					
Heat treatment 80°C (min)	Inoculation temp. (°C)	0.99	0.97	0.94	0.99	0.97	0.94
0	30	6.0	6.3	6.0	3.1	2.0	1.0
	10	5.5	5.5	3.6	2.6	2.2	<1.0
10	30	6.0	6.1	—[a]	2.4	1.3	<1.0
	10	4.6	4.5	<2.0	<1.0	<1.0	<1.0

[a] Not investigated.

From Farkas, J., Beczner, J., and Incze, K., in *Radiation Preservation of Food,* International Atomic Energy Agency, Vienna, 1973, 389. With permission.

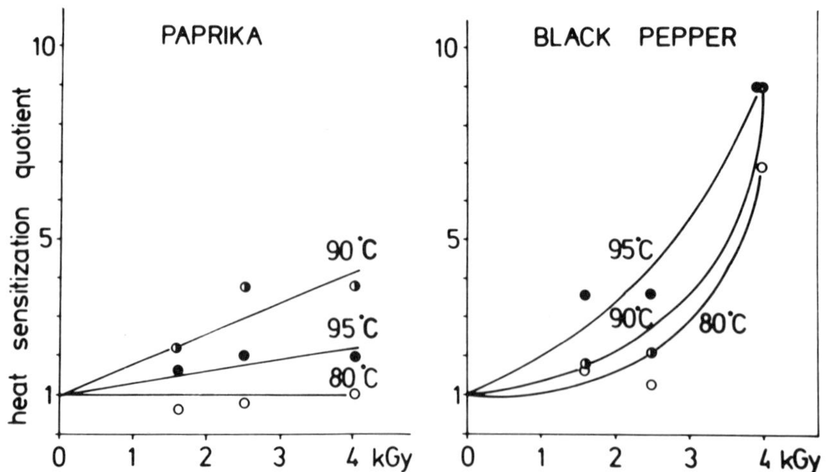

FIGURE 6. The heat sensitization quotients (see text) at various temperature levels, as a function of the irradiation of spice samples. (From Kiss, I. and Farkas, J., *Combination Processes in Food Irradiation,* International Atomic Energy Agency, Vienna, 1981, 112. With permission.)

Since most of the seasoned products of the food industry are undergoing various kinds of processing with some antimicrobial effect, the above sensitization by prior irradiation that appeared in spices may assure a practically complete elimination or effective inhibition of the spice-borne components of their microbial contamination.

The above observations are in good agreement with numerous investigations carried out with pure cultures of microorganisms which prove that, for example, the bacterial spores surviving irradiation become more sensitive to heat and reduced water activity of the growth medium.[260,264-266]

The heat sensitizing effect of irradiation increases with increasing radiation doses. This is illustrated in Figure 6 taken from studies on paprika and black pepper, where the "heat sensitization quotient" represents the ratio of heat destruction times of unirradiated samples to heat destruction times of irradiated samples at given doses and postirradiation heating temperatures in aqueous suspensions of the spices.[114]

The residual viable counts in irradiated black pepper did not increase significantly during a 6-month period, either at a 10° or at 25°C storage temperature, and the increased sensitivity of the residual flora in the irradiated spice was maintained, i.e., no significant repair of the damage induced during the aerobic irradiation conditions occurred during the postirradiation storage.[263] These and similar investigations suggest that the sensitization of the surviving microflora of irradiated dry ingredients is a permanent feature and does not diminish upon a regular storage of the products.[77,262,263] In comparative investigations with black pepper, EtO did not practically affect the heat resistance of the survivors.[77,263]

The differences in the sensitizing effects of irradiation and fumigation make it probable that the nature of injury to bacterial spores differs upon irradiation from that upon fumigation and that radiation treatment is more damaging than EtO to the spore component(s) maintaining the heat resistance of spores. Assuming the role of the radiation-induced damage of cortex peptidoglycan in a partial rehydration of the spore core, and, thereby, in heat sensitization of spores, the inefficiency of EtO to heat sensitize the surviving spores in black pepper makes it probable that EtO is much less destructive to the spore cortex than γ-radiation.

Based on the synergistic effect of ionizing radiation and heat on bacterial spores, Dutch scientists obtained good microbiological results using medium-dose γ-radiation (2 kGy) plus IR treatment designed as a HTST process for decontamination of spices and other dry products.[267]

VIII. STORAGE STABILITY OF IRRADIATED DRY INGREDIENTS

A. Chemical Aspects

Chemical stability of stored spices seems not to be affected by radiation decontamination. The retention of benzene-soluble pigments in ground paprika and changes in surface color during storage were not impaired by a dose of 5 kGy.[74,111,115,313] Szabad and Kiss[268] reported that neither EtO (600 g EtO per cubic meter at 25°C for 6 hr) nor irradiation (5, 9.3, and 11 kGy, respectively) affected the pigment of paprika or the rate of pigment loss upon a 6-month storage. Byun et al.[133] also reported that the acetone-extractable pigment content of hot red pepper powder decreased at the same rate both in untreated and irradiated samples (doses up to 20 kGy) during a 3-month storage.

The loss of volatile oil content in allspice, black pepper, cumin, coriander, marjoram, and nutmeg was not influenced by radiation treatment during several months of storage as compared to the untreated samples.[123,126]

Purwanto et al.[128] reported even somewhat better retention of volatile oil content in irradiated nutmegs than in the untreated ones.

The reducing sugar showed an increasing, and the content of amino-nitrogen a decreasing, tendency with increasing irradiation dose in ginseng powder during storage.[155]

Tipples and Norris[269] observed that the lipid component of irradiated wheat flour oxidized less in 6 months of storage than that of nonirradiated flour.

No significant difference of taste and flavor was found between untreated and 4-kGy treated onion powder during 12-months storage after irradiation.[83]

B. Microbiological Aspects

From the point of view of microbial stability, the mold contamination of spices, herbs, and dry vegetables is of importance. Fungal spoilage of these commodities may occur during storage and shipping, especially at high relative humidity and temperature or if localized wetting occurs. Thus, the ability of surviving molds to propagate in irradiated ground paprika samples was investigated by Farkas et al.[271] and Farkas.[272] Figure 7 shows the counts of

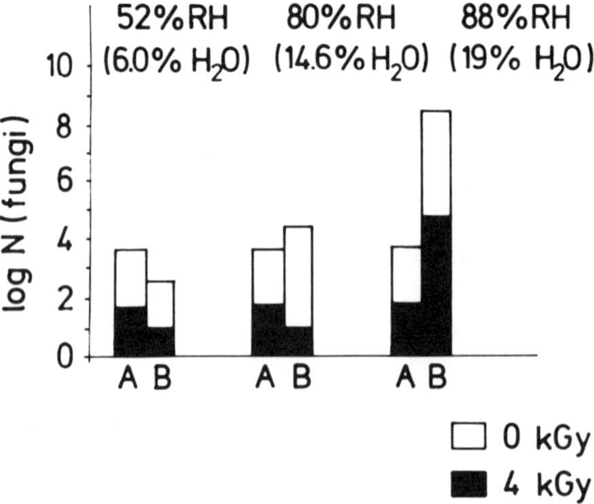

FIGURE 7. Viable mold counts of paprika powder as affected by the relative humidity of the storage space (A) immediately upon radiation treatment and (B) after 4-months storage subsequent to radiation (From Farkas, J., *Aspects of the Introduction of Food Irradiation on Developing Countries,* International Atomic Energy Agency, Vienna, 1973, 43. With permission.)

mold propagules immediately upon radiation treatment of 4 kGy and after 4-months storage at 25°C. As seen in the figure, the dose of 4 kGy reduced the mold count by 99%. The surviving molds in the irradiated paprika were not capable of growth even at 80% relative humidity, and the mold count of radiation-pasteurized paprika powder when stored at 88% relative humidity was still less than 10^5/g after 4 months of storage, while the unirradiated samples when stored under identical conditions were completely molded within a few weeks. In another storage experiment ground paprika samples were stored for 120 days at 0° and 5°C and a relative humidity higher than 90%.[115] The control paprika faded completely and became visibly moldy, while the sample irradiated with 5 kGy remained sound and of bright color, and its residual mold count did not increase during storage even in a highly humid storage space.

In Japanese experiments mold counts of untreated spices (turmeric, rosemary, and white pepper) in polyethylene pouches increased up to 10^8/g during 1 to 3 months of storage at 30° to 35°C and at a humidity above 80% while samples given a 4 kGy irradiation were free from molds.[18] Further studies with additional spices, i.e., black pepper, bay leaves, basil, and thyme, confirmed that 4- to 6-kGy irradiation can suppress the growth of molds in spices packed in polyethylene bags under unfavorable conditions of storage and shipment.[331]

These results can be explained by the observations that surviving microorganisms are damaged and are not capable of reproduction in the medium in which they were exposed to radiation treatment, even if storage conditions were favorable to the growth of undamaged microorganisms (see Chapter 1).

The conclusion is that radiation decontamination doses, if recontamination is excluded by proper packaging and handling, could prevent growth of molds, including eventual toxigenic ones, even under high relative humidity. This points to an important additional potential benefit, particularly for spice-producing developing countries with humid climates.

In many cases a considerably greater decrease occurred in the residual viable cell counts of various irradiated dry ingredients during postirradiation storage than in respective untreated samples.[73,225,249,254,256,271,272] This might be due to somewhat accelerated autooxidation processes in very low moisture aerobic systems.

C. Disinfesting Effect of Radiation Treatment

Some dry ingredients may be infested with insects, either as consequence of contamination at their source or through infestation during storage. Although this treatise does not deal with radiation disinfestation in detail, noting that insects are much less radioresistant than microorganisms, it goes without saying that radiation treatment for microbicidal purposes always involves insect disinfestation as well. The dose requirement for disinfestation is only a fraction of that required for decontamination (disinfection), i.e., it does not exceed 1 kGy.[104,105,273-275] Furthermore, doses inducing sexual sterility of insects are much lower than those required to cause immediate mortality (see also this chapter, Section III.B).

Packaging which cannot be penetrated by insects is, of course, a crucial requirement for the success of insect-free storage of disinfested products.[304]

IX. EXPERIMENTS ON THE USE OF IRRADIATED INGREDIENTS IN FOOD PROCESSING

In a Russian study introduction of radiation-sterilized spice mixtures into sprat preserves resulted in a considerable reduction in spoilage and in an increase of shelf-life from 14 to 20 days.[276]

In Hungarian experiments an additive mixture used in the manufacturing of canned pork liver paste, consisting of ground rice, common salt, ''French seasoning'', and powdered onion, was irradiated at 15 kGy, and comparative canning studies were performed with various F_0 values using either untreated or irradiated ingredients.[272] The total viable aerobic cell count of the product prior to the canning heat treatment was 7×10^7 per can, of which the total aerobic cell count introduced by the untreated seasoning amounted to 2.9×10^6 per can, i.e., about 4% of the aerobic microbial contamination. In spite of this, the results showed the effectiveness of the canning heat treatment to depend to a great extent on the microbial contamination of the additive mixture. The use of radappertized additives resulted in a substantial reduction of the heat treatment requirement of canning with unchanged microbiological safety. Similar canning trials on the production of canned minced meat with 5-kGy-irradiated dry ingredients (wheat flour, sodium caseinate, and spices) also gave encouraging results in reducing the heat treatment.[109] By reducing the extent of heat treatment, the quality (taste, color, and consistency) can be improved and the energy consumption and processing time can be reduced. By the latter, the production capacity will increase the heat treatment requirement and improve product quality.[109]

In another experiment in the meat processing industry, viable cell counts in brawn loaves (head cheese) manufactured with irradiated seasonings were substantially lower after cooking than those prepared with untreated seasonings.[74] The difference in viable cell count of brawn samples made with seasonings treated with 3 or 15 kGy was not significant. It was found that the use of irradiated seasonings reduced the heat treatment requirement of the brawn by about one half, while the microbiological quality of the product was improved.

Similar results were reported by German authors who made a comparative study of the production of pasteurized tinned pork seasoned with irradiated spice mixtures (7.5 and 10 kGy) and untreated spices, respectively (see Table 32).[277]

Table 32
MICROBIAL STABILITY OF PASTEURIZED TINNED PORK (HAM) AS AFFECTED BY THE RADIATION DECONTAMINATION OF ITS SEASONING SPICE MIXTURE[277]

Storage temp. (°C)	Shelf-life (days) of batches prepared with spices irradiated at (kGy)		
	0	7.5	10
8	≥180	>180	>180
15	96	>180	>180
20	15	~30	>90

REFERENCES

1. **Dunn, G. G., Campbell, W. L., Fram, H., and Hutchins, A.,** Biological and photochemical effects of high energy electrostatically produced roentgen rays and cathode rays, *J. Appl. Phys.,* 19, 605, 1948
2. **Proctor, B. E., Goldblith, S. A., and Fram, H.,** Effect of supervoltage cathode rays on bacterial flora of spices and other dry food materials, *Food Res.,* 15, 490, 1950.
3. International Trade Centre UNCTAD/GATT, *Spices, a Survey of the World Market,* Vol. 1 and 2, ITC, Geneva, 1982.
4. **Kammer, J.,** Über die besonderen Eigenschaften der aeroben Bazillen in Gewürzen unter besonderer Berücksichtigung der Eiweisszersetzung, Veterinary medicine dissertation, University of Hannover, Hannover, West Germany, 1961.
5. **Eschmann, K. H.,** Gewürze — eine Quelle bakteriologischer Infektionen, *Alimenta,* 4(3), 83, 1965.
6. **Fievez, L. and Granville, A.,** Alterations de semiconserves de viande par des épiches riches en spores bactériennes tres résistantes á la chaleur, *Ann. Med. Vet.,* 109, 143, 1965.
7. **Elter, B. and Scharner, E.,** Über den Keimgehalt von Gewürzen und seine Bedeutung für die Fleischwarenproduktion, *Fleisch,* 22, 219, 1968.
8. **Heath, H. B.,** Bacteria free spices in food processing, *Food Process. Packag.,* 4, 144, 1964.
9. **Julseth, R. M. and Deibel, R. H.,** Microbial profile of selected spices and herbs at import, *J. Milk Food Technol.,* 37, 414, 1974.
10. International Commission on Microbiological Specifications for Foods, Spices, in *Microbial Ecology of Foods,* Vol. 2, Silliker, J. H., Elliott, R. P., Baird-Parker, A. C., Bryan, F. L., Christian, J. H. B., Clark, D. S., Olson, J. C., Jr., and Roberts, T. A., Eds., Academic Press, New York, 1980, 731.
11. **Glas, A.,** *Praktisches Handbuch der Lebensmittel,* Bayerischer Landwirtschaftsverlag, Munich, 1965.
12. **Soedarman, H., Stegeman, H., Farkas, J., and Mossel, D. A. A.,** Decontamination of black pepper by gamma radiation, in *Microbial Associations and Interactions in Food,* Kiss, I., Deák, T., and Incze, K., Eds., Akadémiai Kiadó, Budapest, 1984, 401.
13. **Palumbo, S. A., Rivenburgh, A. I., Smith, J. L., and Kissinger, J. C.,** Identification of *Bacillus subtilis* from sausage products and spices, *J. Appl. Bacteriol.,* 38, 99, 1975.
14. **Inal, T.,** Untersuchungen über die Bakterienflora der Gewürze unter besonderer Berücksichtigung der aeroben sporenbilder, *Veteriner Fakultesi Clergisi,* 12(1—2), 82, 1965.
15. **Goto, A., Yamazaki, K., and Oka, M.,** Bacteriology of radiation sterilization of spices, *Food Irradiat. Jpn.,* 6(1), 35, 1971.
16. **Seenappa, M. and Kempton, A. G.,** A note on the occurrence of *Bacillus cereus* and other species of *Bacillus* in Indian spices of export quality, *J. Appl. Bacteriol.,* 50, 225, 1981.
17. **Fabri, I., Nagel, V., Tabajdi-Pinter, V., Zalavari, Zs., Szabad, J., and Deak, T.,** Qualitative and quantitative analysis of aerobic spore-forming bacteria in Hungarian paprika, in *Fundamental and Applied Aspects of Bacterial Spores,* Dring, G. J., Ellar, D. J., and Gould, G. W., Eds., Academic Press, London, 1985, 455.
18. **Ito, H., Watanabe, H., Bagiawati, S., Muhamad, L. J., and Tamura, N.,** Distribution of microorganisms in spices and their decontamination by gamma-irradiation, Paper IAEA-SM-271/110 p, at IAEA/FAO Int. Symp. on Food Irradiat. Proc., Washington, D.C., March 4 to 8, 1985.

19. **Neumayr, L., Promeuschel, S., Arnold, I., and Leistner, L.,** *Gewürzentkeimung, Verfahren und Notwendigkeit,* Abschlussbericht für die Adalbert-Raps-Stiftung zum Forschungsvorhaben, Institut für Mikrobiologie, Toxikologie und Histologie der Bundesanstalt für Fleischforschung, Kulmbach, 1983.
20. **Pohja, M. S.,** Vergleichende Untersuchungen über den Mikrobengehalt fester und flüssiger Gewürze, *Fleischwirtschaft,* 9, 547, 1957.
21. **Richmond, B. and Fields, M. L.,** Distribution of thermophilic aerobic sporeforming bacteria in food ingredients, *Appl. Microbiol.,* 14, 623, 1966.
22. **Pruthi, J. S.,** Spices and condiments, in *Chemistry, Microbiology, Technology,* Academic Press, New York, 1983.
23. **Sheneman, J. N.,** Microbiology of dehydrated onion products. I. Survey of aerobic mesophilic bacteria, *J. Food Sci.,* 38, 206, 1973.
24. **Michels, M. J. and Visser, F. M. W.,** Occurrence and thermoresistance of spores of psychrophilic and psychrotrophic aerobic spore formers in soil and food, *J. Appl. Bacteriol.,* 41, 1, 1976.
25. **deBoer, E. and Janssen, F. W.,** Microbiology of spices and herbs (in Dutch), presented at Dutch Symp. on Food Microbiology, Delft, November 17, 1983.
26. **Powers, E. M., Latt, T. G., and Brown, T.,** Incidence and levels of *Bacillus cereus* in processed spices, *J. Milk Food Technol.,* 39, 668, 1976.
27. **Kim, H. K. and Goepfert, J. M.,** Occurrence of *Bacillus cereus* in selected dry food products, *J. Milk Food Technol.,* 34, 12, 1971.
28. **Volkova, R. S.,** *Bacillus cereus* contamination of foods and environment at institutional feeding points (in Russian), *Gig. Sanit.,* 36, 108, 1970.
29. **Baxter, R. and Holzapfel, W. H.,** A microbial investigation of selected spices, herbs, and additives in South Africa, *J. Food Sci.,* 47, 570, 1982.
30. **Smith, L. D. S.,** *Clostridium perfringens* food poisoning, in *Microbiological Quality of Foods,* Slanetz, L. W., Chichester, C. O., Gaufin, A. R., and Ordal, Z. J., Eds., Academic Press, New York, 1963, 77.
31. **Strong, D. H., Canada, J. C., and Griffiths, B. B.,** Incidence of *Clostridium perfringens* in American foods, *Appl. Microbiol.,* 11, 42, 1963.
32. **Nikolaeva, S. A.,** *Cl. perfringens* in foodstuffs and in semimanufactured products in the canning industry (in Russian), *Gig. Sanit.,* 32(5), 30, 1976.
33. **Krishnaswamy, M. A., Patel, J. D., and Parthasarathy, N.,** Enumeration of microorganisms in spices and spice mixtures, *J. Food Sci. Technol.,* 8, 191, 1971.
34. **Powers, E. M., Lawyer, R., and Masuoka, Y.,** Microbiology of processed spices, *J. Milk Food Technol.,* 38, 683, 1975.
35. **El-Mossalani, E. and Youssef, A.,** Studies on bacterial contaminaton of spices used in meat products, *Zentralbl. Veterinaermed. Reihe B,* 12, 176, 1965.
36. **Karlsson, K. E. and Gunderson, M. F.,** Microbiology of dehydrated soups. II. Adding machine approach, *Food Technol.,* 19, 86, 1965.
37. **Kadis, V. W., Hill, D. A., and Pennifold, K. S.,** Bacterial content of gravy bases and gravies obtained in restaurants, *Can. Inst. Food Technol. J.,* 4, 130, 1971.
38. **Schwab, A. H., Harpestad, A. D., Schwarzentruber, A., Lanier, J. M., Wentz, B. A., Duran, A. P., Barnard, R. J., and Read, R. B., Jr.,** Microbiological quality of some spices and herbs in retail markets, *Appl. Environ. Microbiol.,* 44, 627, 1982.
39. **Christensen, C. M., Fanse, H. A., Nelson, G. H., Bates, F., and Mirocha, C. J.,** Microflora of black and red pepper, *Appl. Microbiol.,* 15, 622, 1967.
40. **Krishnaswamy, M. A., Nair, K. K. S., Patel, J. D., and Parthasarathy, N.,** Preliminary observations on the survival of *Salmonella* in curry, samber, coriander and red-chilli powders, *J. Food Sci. Technol.,* 12, 195, 1975.
41. **Wilson, C. R. and Andrews, W. H.,** Sulfite compounds as neutralizers of spice toxicity for *Salmonella, J. Milk Food Technol.,* 39, 464, 1976.
42. **Sperber, W. H. and Deibel, R. H.,** Accelerated procedure for *Salmonella* detection in dried foods and feeds involving only broth cultures and serological reactions, *Appl. Microbiol.,* 17, 533, 1969.
43. **Laidley, R., Handzel, S., Severs, D., and Butler, R.,** *Salmonella weltevreden* outbreak associated with contaminated pepper, *Epidemiol. Bull. (Dep. Natl. Health Welfare, Ottawa),* 18(4), 62, 1974.
44. WHO, *Wkly. Epidemol. Rec. World Health Organ. Geneva,* 48, 377, 1973.
45. WHO, *Salmonella* surveillance, *Wkly. Epidemol. Rec. World Health Organ. Geneva,* 42, 351, 1974.
46. WHO, Foodborne disease surveillance, outbreak of *Salmonella oranienburg* infection, *Wkly. Epidemol. Rec. World Health Organ. Geneva,* 57, 329, 1982.
47. **Masson, A.,** Hygienic quality of spices, *Mitt. Geb. Lebensmittelunters. Hyg.,* 69, 544, 1978.
48. **Hadlok, R.,** Schimmelpilzkontamination von Fleischerzeugnissen durch naturbelassenene Gewürze, *Fleischwirtschaft,* 49, 1601, 1969.
49. **Flannigan, B. and Hui, S. C.,** The occurrence of aflatoxin-producing strains of *Aspergillus flavus* in the mold floras of ground spices, *J. Appl. Bacteriol.,* 41, 411, 1976.

50. **Horie, Y., Samazaki, M., Miyaki, K., and Udagawa, S.,** On the fungal contents of spices, *J. Food Hyg. Soc. Jpn.*, 12, 516, 1971.
51. **Pal, N. and Kundu, A. K.,** Studies on *Aspergillus* ssp. from Indian spices in relation to aflatoxin production, Sci. Cult., 30, 252, 1972.
52. **Moreno-Martinez, E. and Christensen, C. M.,** Fungus flora of black and white pepper (*Piper nigrum* L.), *Rev. Latinoam. Microbiol.*, 15, 19, 1973.
53. **Dragoni, I.,** Role of fungal contamination of black pepper in mould growth on dry sausages, *Ind. Aliment. (Pinerolo, Italy)*, 17, 280, 1978.
54. **Moreau, C. and Moreau, M.,** La contamination des épices, ses conséquences dans les industries alimentaires, *Ind. Aliment. Agric.*, 95, 497, 1978.
55. **Eckhardt, Ch. and Leistner, L.,** unpublished results; see also Reference 19.
56. **Vaugh, R. H.,** The microbiology of dehydrated vegetables, *Food Res.*, 16, 429, 1951.
57. **Christensen, C. M.,** Pure spices — how pure?, *Am. Soc. Microbiol. News*, 38, 165, 1972.
58. **Shank, R. C., Wogan, G. N., and Gibson, J. F.,** Dietary aflatoxins and human liver cancer. I. Toxigenic moulds in foods and foodstuffs of tropical South East Asia, *Food Cosmet. Toxicol.*, 10(1), 51, 1972.
59. **Scott, P. M. and Kennedy, B. P. C.,** Analysis and survey of ground black, white and capsicum peppers for aflatoxins, *J. Assoc. Off. Agric. Chem.*, 56, 1452, 1973.
60. **Suzuki, J. I., Daninius, B., and Kilbuck, J. H.,** A modified method for aflatoxin determination in spices, *J. Food Sci.*, 38, 949, 1973.
61. **Scott, P. M. and Kennedy, B. P. C.,** The analysis of spices and herbs for aflatoxins, *Can. Inst. Food Technol. J.*, 8, 124, 1975.
62. **Llewellyn, G. C., Burkett, M. L., and Eadie, T.,** Potential mold growth, aflatoxin production and antimycotic activity of selected natural spices and herbs, *J. Assoc. Off. Anal. Chem.*, 64, 955, 1981.
63. **Hill, M. H.,** Seasoning in meat products, *Process Biochem.*, 6(12), 27, 1971.
64. **Schönberg, F.,** Zur Verwendung von Gewürzessenzen aus Naturgewüezen für die Herstellung von Fleischwaren, *Fleischwirtschaft*, 14, 272, 1962.
65. **Bartels, H. B. and Hadlok, R.,** Organoleptische und bakteriologische Untersuchungen von Gewürzextrakten aus Naturgewürzen, *Fleischwirtschaft*, 46, 234, 1966.
66. **Pekhov, A. V., Katyuzhanskaya, A. N., and Dyobankova, N. F.,** Ispol'zovanie ekstraktov pryanuh rastenii v proizvodstve konservov (The use of seasoning extracts in the canning industry) (in Russian), *Konservn. Ovoshchesush. Prom.*, 7, 12, 1969.
67. **Schurm, H.,** Seasoning hard sausages with spice extracts, *Food Trade Rev.*, 41(3), 31, 1969.
68. **Wyler, O. D.,** Hilfs- und Zusatzstoffe und ihr Einfluss auf Technologie and Qualität von Fleischwaren, *Fleischwirtschaft*, 52, 727, 1972.
69. **Körmendy, L.,** Különbözö füszerfélék és füszerolajok vizsgálata. I. Bors és borskivonat összehasonlitása (Examination of domestic and foreign spices and spice oils. I. Comparing pepper and pepper extract) (in Hungarian), *Husipar*, 22, 123, 1973.
70. **Weber, H. and Gährs, H.,** CO_2-Hochdruckextraktion — ein neues Verfahren zur Gewürzextraktgewinnung, *Fleischwirtschaft*, 63, 1747, 1983.
71. *Begriffbestimmungen für Gewürze, Ersatzgewürze und daraus hergestellte Erzeugnisse*, Schriftenreihe des Bundes für Lebensmittelrecht und Lebensmittelkunde, Heft 51, B. Behr's Verlag, Hamburg, 1964.
72. **Hadlok, R. and Toure, B.,** Mykologische und bakteriologische Untersuchungen entkeimter Gewürze, *Arch. Lebensmittelhyg.*, 24(1), 20, 1973.
73. **Bachman, S. and Gieszczynska, J.,** Studies on some microbiological and chemical aspects of irradiated spices, in *Aspects of the Introduction of Food Irradiation in Developing Countries*, International Atomic Energy Agency, Vienna, 1973, 33.
74. **Farkas, J., Beczner, J., and Incze, K.,** Feasibility of irradiation of spices with special reference to paprika, in *Radiation Preservation of Food*, International Agency for Atomic Energy, Vienna, 1973, 389.
75. **Farkas, J. and Andrássy, É.,** Investigations of the surviving microflora in irradiated black pepper (in Hungarian), *Husipar*, 32(3), 113, 1983.
76. **Farkas, J. and Andrássy, É.,** Comparative investigation of some effects of gamma radiation and ethylene oxide on aerobic bacterial spores in black pepper. Abstracts of papers at satellite meeting: Spore Radiobiology, 7th Int. Congr. of Radiat. Res., *Int. J. Radiat. Biol.*, 45, 411, 1984.
77. **Farkas, J. and Andrássy, É.,** Comparative investigation of some effects of gamma radiation and ethylene oxide on aerobic bacterial spores in black pepper, in *Proc. of IUMS-ICFMH 12th Int. Symp.: Microbial Associations and Interactions in Food*, Kiss, I., Deák, T., and Incze, K., Eds., Akadémiai Kiadó, Budapest, 1984, 393.
78. **Briggs, A.,** The resistance of spores of the genus *Bacillus* to phenol, heat and radiation, *J. Appl. Bacteriol.*, 29, 490, 1960.
79. **Härnulv, B. G. and Snygg, B. G.,** Radiation resistance of spores of *Bacillus subtilis* and *B. stearothermophilus* at various water activities, *J. Appl. Bacteriol.*, 36, 677, 1973.

80. **Török, G. and Farkas, J.**, Investigations into the reduction of viable cell count of paprika powder by ionizing radiation (in Hungarian), *Comm. Central Food Res. Inst. Budapest,* 3, 1, 1961.
81. **Coretti, K.**, Kaltenkeimung von Gewürzen mit Athylenoxid, *Fleischwirtschaft,* 9, 183, 1957.
82. **Mayr, G. E. and Suhr, H.**, Preservation and sterilization of pure and mixed spices, in *Conf. Proc. Spices, Trop. Prod. Inst. London,* April 1972, 201.
83. **Eiss, M. I.**, Irradiation of spices and herbs, *Food Technol. Aust.*, 36, 362, 1984.
84. **Vajdi, M. and Pereira, N. N.**, Comparative effects of ethylene oxide, gamma irradiation and microwave treatment on selected spices, *J. Food Sci.*, 38, 893, 1973.
85. **Franse, O.**, personal communication, 1984.
86. **Kiss, I., Beczner, J., Kovács, E., and Kovács, S.**, Commercial Scale Irradiation Experiments with Spices, Vegetables and Fruits, 3rd Progress Report, IAEA Research Agreement No. 3033/CF, Central Food Research Institute, Budapest, December 1983.
87. **Lerke, P. A. and Farber, L.**, Effect of electron beam irradiation on the microbial content of spices and teas, *Food Technol.*, 14, 266, 1960.
88. **Josimovic, L.**, Radiation Treatment of Spices and Some Protein Rich Food, IAEA Research Contract No. 3277/RB, January 1983—September 1983, Boris Kidric Institute of Nuclear Sciences, Vinca, Yugoslavia, 1983.
89. **van Dijck, J. G. M.**, Study on the Application Possibilities of Irradiation for Decontamination of Spices (in Dutch), Rapport No. 3, Proefbedrijf Voedselbestraling, Wageningen, 1970.
90. **Gerhardt, U. and Ladd-Effio, J. C.**, Entkeimung von Gewürzen durch Athylenoxid-Begasung, *Dtsch. Molk. Ztg.*, 103, 1016, 1982.
91. **Theivendirarajah, K. and Jajewardene, A. L.**, The effect if gamma irradiation (^{60}Co) on spices and red onion, IAEA res. contract, no. 2840/JN, in FAO/IAEA Res. Coordination Meet. on the Asian Regional Cooperative Project in Food Irradiation, Bangkok, November 22 to 26, 1982.
92. **Funke, D., Krey, P., and Jantz, A.**, Microbiological and Sensory Evaluation of Several γ-Irradiated Spices, ZFI-Mitteilungen No. 98, Zentralinstitut für Isotopen- und Strahlenforschung, Leipzig, 1984, 642.
93. **Dirkse, H.**, personal communication, 1982.
94. **Zehnder, H. J. and Ettel, N.**, Zur Keimzahlverminderung in Gewürzen mit Hilfe ionisierender Strahlen. III. Mittl.: Mikrobiologische, sensorische und physikalisch-chemische Untersuchungen verschiedener Gewürze, *Alimenta,* 20, 95, 1982.
95. **Sharma, A., Ghanekar, A. S., Padwal-Desai, S. R., and Nadkarni, G. B.**, Microbiological status and antifungal properties of irradiated spices, *J. Agric. Food Chem.*, 32, 1061, 1984.
96. **Weber, H.**, Gewürzentkeimung, Einflüsse von Elektronen und Gammastrahlen auf die Qualität verschiedener Gewürze, *Fleischwirtschaft,* 63, 1065, 1983.
97. **Inal, T., Keskin, S., Tolgay, Z., and Tezcan, J.**, Gewürzsterilization durch Anwendung von Gamma-Strahlen, *Fleischwirtschaft,* 55, 675, 1975.
98. **Grünewald, Th.**, Electron irradiation of dry food products, *Radiat. Phys. Chem.*, 22, 733, 1983.
99. **Grünewald, Th.**, Untersuchungen zur Bestrahlung von Trockenprodukten, BFE-R-84-02, Berichte der Bundesforschungsanstalt für Ernährung, Karlsruhe, July 1984.
100. **Pintauro, N. D.**, *Food Additives to Extend Shelf Life,* National Dairy Council, Rosemont, Ill., 1974, 302.
101. **van Kooij, J. G.**, Neuere Erkentnisse Über die Bestrahlung von Früchten und Gewürzen, in *Nahrungsmittelbestrahlung,* Somogyi, J. C., Ed., Forster Verlag, Zurich, 1967. 44.
102. **Pappas, H. J. and Hall, L. A.**, The control of thermophilic bacteria, *Food Technol.*, 6, 456, 1959.
103. **Makó, I., Mikes, Gy., and Takács, J.**, Sterilization of spices (in Hungarian), *Husipar,* 8, 23, 1959.
104. **Saputra, T. S., Farkas, J., Maha, M., and Purwanto, Z. I.**, Trial Intercountry Shipment of Irradiated Spices, IFFIT Rep. No. 47, International Facility for Food Irradiation Technology, Wageningen, April 1984.
105. **Farkas, J.**, Evaluation of Trial Shipments of Indonesian Spices, Interim Report, International Facility for Food Irradiation Technology, Wageningen, 1983.
106. **Tjaberg, T. B., Underdal, B., and Lunde, G.**, The effect of ionizing radiation on the microbiological content and volatile constituents of spices, *J. Appl. Bacteriol.*, 35, 473, 1972.
107. **Silberstein, O., Galetto, W., and Henzi, W.**, Irradiation of onion powder: effect on microbiology, *J. Food Sci.*, 44, 975, 1979.
108. **Kiss, I. and Farkas, J.**, Extension of the shelf-life of foods by irradiation (in Hungarian), *Élelmez. Ipar,* 37, 81, 1983.
109. **Kiss, I., Zachariev, Gy., Farkas, J., Szabad, J., and Tóth-Pesti, K.**, The use of irradiated ingredients in food processing, in *Food Preservation by Irradiation,* Vol. 1, International Atomic Energy Agency, Vienna, 1979, 263.
110. **Farkas, J. and El-Nawavy, A. S.**, Effect of gamma radiation upon the viable cell count and some other quality characteristics of dried onions, *Acta Aliment.*, 2, 437, 1973.
111. **Zachariev, Gy.**, Improving Pigment Stability of Ground Paprika at Heat- and Radiation Treatment (in Hungarian), Progress Rep. Central Food Research Institute, Budapest, March 1984.

112. **Katušin-Ražem, B., Antolic, M., Ražem, O., Dvornik, I., Briski, B., and Vrabec, A.,** Microbiological decontamination of spices by ionizing radiation (in Croatian), *Tehnol. Mesa,* 24, 115, 1983.
113. **Kiss, I.,** Reduction of microbial contamination of spices by irradiation, in *Proc. 28th Eur. Meet. of Meat Res. Workers,* Vol. 1, Instituto del Frio, Madrid, 1982, 322.
114. **Kiss, I. and Farkas, J.,** Combined effect of gamma irradiation and heat treatment on microflora of spices, in *Combination Processes in Food Irradiation,* International Atomic Energy Agency, Vienna, 1981, 107.
115. **Beczner, J. and Farkas, J.,** Storage experiments with ground seasoning paprika treated by irradiation for reduction of cell count (in Hungarian), *Kiserletugyi Kozl.* E67 (1—2), 3, 1974.
116. **Szabad, J.,** Comparative Studies into the Cell-Count Reducing Effects of Ethylene Oxide Treatment and Gamma Radiation (in Hungarian), Thesis, University of Szeged, 1977.
117. **Zehnder, H. J., Ettel, W., and Jackob, M.,** Zum Keimverminderung bei Gewürzen mit Hilfe ionisierender Strahlen. I. Mitt.: Qualitative Parameter bei schwarzem Pfeffer, *Alimenta,* 18, 43, 1979.
118. **Saputra, T. S., Maha, M., and Purwanto, Z. I.,** Gamma Irradiation of Spices. II. Quality Changes of Irradiated Spices Stored in Different Indigeneous Packaging Materials, Progress Rep. to the International Atomic Energy Agency, Contract No. 2630/R1/IN, Jakarta, December 1981 to September 1982.
119. **Robinson, R. F., Overbeck, R. C., and Porter, F. E.,** X-ray sterilization of spices, *Coffee Tea Ind. Flavor Field,* 77, 61, 1954.
120. **Beczner, J. and Kiss, I.,** Reduction of viable cell counts in spices by irradiation (in Hungarian), *Konzerv. Paprikaip.,* (3), 108, 1983.
121. **Sirnik, M. and Gorisek, M.,** Microflora of spices for the meat industry (in Croatian), *Tehnol. Mesa,* 24, 107, 1983.
122. **Delkinova, S. M. and Dupuy, P.,** Sanitation of onion powder by irradiation and heating, *Acta Aliment.* 8, 205, 1973.
123. Medicinal Plant Research Institute, The Effects of Ionizing Radiation on the Essential Oil Content of Spices, Rep. to the Int. Project in the Field of Food Irradiation, Budakalász and the Central Food Research Institute, Budapest, September 1979.
124. **Sharabash, M. T. M.** Studies on spices irradiation in Egypt, presented at the Food and Agric. Organ./ Int. At. Energy Agency Res. Coordination Meet. on Pre-Commercial Scale Radiat. Treatment of Foods, Manila, December 12 to 16, 1983.
125. **Bachman, S., Witkowski, S., and Zegota, A.,** Some chemical changes in irradiated spices (caraway and cardamom) in *Food Preservation by Irradiation,* Vol. 1, International Atomic Energy Agency, Vienna, 1978, 435.
126. **Hilmy, N., Chosdu, R., Sudiro, F. S., and Syuib, F.,** Studies on gamma irradiated medicinal plants and spices. I. Myristica argentea, Myristica fragrans, Coriandrum sativum and Foeniculum vulgare (in Indonesian), *Malajah BATAN,* 14(3), 37, 1981.
127. **Kuruppu, D. P., Langerak, D. Is., and van Duren, M. D. A.,** Effect of Gamma Irradiation, Fumigation and Storage Time on Volatile Oil Content of Some Spices, IFFIT Rep., No. 41, International Facility for Food Irradiation Technology, Wageningen, 1983.
128. **Purwanto, Z. I., Langerak, D. Is., and van Duren, M. D. A.,** Effect of a Combination of Packaging and Irradiation on the Chemical and Sensoric Quality of Ground Nutmeg during Storage under Tropical Conditions, IFFIT Rep., No. 31, International Facility for Food Irradiation Technology, Wageningen, 1983.
129. **Uchman, W., Fiszer, W., Mróz, I., and Pawlik, A.,** The influence of radappertization upon some sensory properties of black pepper, *Nahrung,* 27, 461, 1983.
130. **Vajdi, M.,** Comparative Effects of Ethylene Oxide, Gamma Irradiation and Microwave Treatments on the Control of Microorganisms in Selected Spices, M. Sc. thesis, University of Manitoba, Winnipeg, October 1970.
131. **Bahari, I., Ishak, S., and Ayub, M. K.,** The effect of gamma radiation and storage time of the volatile constituents, piperine, piperettine and sensory quality of pepper, *Sains Nucl.* 1(3), 1, 1983.
132. **Zehnder, H. J.,** Zur Keimzahlverminderung bei Gewürzen mit Hilfe ionisierender Strahlen. II. Metteilung: Beeinflussung der Schärfe von schwarzem Pfeffer, *Alimenta,* 19, 17, 1980.
133. **Byun, M. W., Kwon, J. H., and Cho, H. O.,** Sterilization and storage of spices by irradiation. I. Sterilization of powdered hot pepper paste (in Korean), *Korean J. Food Sci. Technol.,* 15, 359, 1983.
134. **Beczner, J., Farkas, J., Watterich, A., Buda, B., and Kiss, I.,** Study into the identification of irradiated ground paprika, in *Proc. Int. Colloq. Identification of Irradiat. Foodstuffs,* Commission of the European Communities, Directorate — General Scientific and Technical Information and Information Management, Luxembourg, 1974, 255.
135. **Varsányi, I., Liptay, G., Farkas, J., and Petrik-Brandt, E.,** Thermal analysis of spices decontaminated by irradiation, *Acta Aliment.* 8, 397, 1979.
136. **Galetto, W., Kahan, J., Eiss, M., Welbourn, J., Bednarczyk, A. K. and Silberstein, D.,** Irradiation treatment of onion powder: effects on chemical constituents, *J. Food Sci.,* 44, 591, 1979.

137. **Kiss, I., Farkas, J., Ferenczi, S., Kálmán, B., and Beczner, J.,** Effects of irradiation on the technological and hygienic qualities of several food products, in *Improvements of Food Quality by Irradiation,* STI/PUB/ 370, International Atomic Energy Agency, Vienna, 1974, 158.
138. **Josimovic, Lj.,** Study on some chemical changes in irradiated pepper and parsley, *Int. J. Appl. Radiat. Isot.,* 34, 787, 1983.
139. Netherlands Spice Trade Association, unpublished studies, 1984.
140. **Hall, R. L.,** Effect of Ionizing Radiation on Spices, Rep. I-VIII. U.S. Army Quartermaster Corps, Contract 7-84-01-002, N-28, McCormick Company, Inc., Hunt Valley, Md., 1955 to 1958.
141. **Funke, D., Krey, P., and Jantz, A.,** Microbiological and Sensory Evaluation of Several Gamma-Irradiated Spices, ZFI-Mitteilungen No. 98, Zentralinstitut für Isotopenund Strahlenforschung, Leipzig, 1984, 642.
142. **Saputra, T. S., Harsoyo, A., and Sudarman, H.,** Gamma irradiation of spices, IAEA-SR-60, presented at the Food and Agric. Organ./Int. At. Energy Agency Semin. on Food Irradiat. for Developing Countries in Asia and the Pacific, Tokyo, November 9 to 13, 1981.
143. **Hansen, P. -I. E.,** Radiation treatment of meat products and animal by-products, in *Food Irradiation,* International Atomic Energy Agency, Vienna, 1966, 411.
144. **Kiss, I.,** Pre-commerical scale irradiation experiments with spices, vegtables and fruits, *Food Irradiat. Newsl.,* 8 (1), 10, 1984.
145. **Silberstein, O., Kahan, J., Penniman, J., and Henzi, W.,** Irradiation on onion powder; effects on taste and aroma characteristics, *J. Food Sci.,* 44, 971, 1979.
146. **Byun, M. W., Kwon, J. H., and Cho, H. O.,** Sterilization and storage of onion powder by irradiation (in Korean), *Korean J. Food Sci. Technol.,* 16(1), 47, 1984.
147. **Kuruppu, D. P., Schmidt, K., Langerak, D. Is., van Duren, M. D. A., and Farkas, J.,** Effects of Irradiation and Fumigation on the Antioxidative Properties of Some Spices. *Acta Aliment.,* 14, 343, 1985.
148. Markets for Selected Medicinal Plants and their Derivatives, International Trade Centre UNCTAD/GATT, ITC, Geneva, 1982.
149. **Katušin-Ražem, B., Ražem, D., Dvornik, I., and Matic, S.,** Radiation treatment of herb tea for reduction of microbial contamination *(Flores chamomillae), Radiat. Phys. Chem.,* 22, 707, 1983.
150. **Thery, V.,** Qualite des plantes a infusion de la recolte a la commercialisation apport potential due traitment ionisant, Thesis, Institute National Agronomique, Paris-Grignon/Centre d'Etudes Nucleaires de Cadarache, Saint Paul Lez Durance, France, 1984.
151. **Devleeschouwer, M. J. and Dony, J.,** Normes microbiologiques des drogues d'origine végétale et leurs mélanges, *J. Pharm. Belg.,* 34, 260, 1979.
152. **Saint-Lebe, L., Henon, Y., and Thery, V.,** Le traitement ionisant des produits secs et deshydrates: cas des plantes medicinales a infusion, IAEA-SM-271/12, presented at the Int. At. Energy Agency/Food and Agric. Organ. Symp. on Radiat. Processing of Food, Washington, D.C., March 4 to 8, 1985.
153. Petition for Authorization on the Application of Ionizing Radiations for Decontamination of Infusions (in French), L'Institut National des Radioelements, Fleurus, Belgium, 1983.
154. **Katušin-Ražem, B., Ražem, D., Dvornik, I., Matic, S., and Mihokovic, V.,** Radiation Decontamination of Dry Chamomile Flowers and Chamomile Extract, IAEA-SM-271/75, presented at the Int. At. Energy Agency/Food and Agric. Organ. Symp. on Food Irradiat. Processing, Washington, D.C., March 4 to 8, 1985.
155. **Min, O. N.,** A study on the storage of ginseng powder by gamma-irradiation (in Korean) *Korean J. Environ. Health Soc.,* 8(1), 45, 1982.
156. **Sung, H. S., Park, M. H., Lee, K. S., and Cho, H. O.,** Studies on the preservation of Korean ginseng by irradiation (in Korean), *J. Korean Agric. Chem. Soc.,* 25(2), 105, 1982.
157. **Michels, M. J. M.** Die mikrobiologische Qualität von Trockengemüse, *Z. Lebensm. Technol. Verfahrenstechnik,* 29(1), 14, 1978.
158. **Clark, W. S., Reinbold, G. W., and Rambo, R. S.** Enterococci and coliforms in dehydrated vegetables, *Food Technol.,* 20, 1353, 1966.
159. The International Commission on Microbiological Specifications for Foods, Vegetables, Fruits, Nuts, and their Products, in *Microbial Ecology of Foods,* Vol. 2, Silliker, J. H., Elliot, R. P., Baird-Parker, A. C., Bryan, F. L., Christian, J. H. B., Clark, D. S., Olson, J. C., Jr., and Roberts, T. A., Eds., Academic Press, New York, 1980, 606.
160. **Thoman, R., Loest, K., and Kretschmer, P.,** Results of the Investigations with Irradiated Spices, ZFI-Mitteilungen No. 98, Zentralinstitut für Isotopen- und Strahlenforschung, Leipzig, 1984, 636.
161. **Schroeder, C. W.,** Dehydrating Vegetables, U.S. Patent 3025171, 1962.
162. Unilever Ltd., British Patent 874 752/4. 7. 1958.
163. **Farkas, J., Kálmán, B., Bencze-Böcs, J., and Jauernig, A.,** Experiments on improving the quality of dehydrated vegetables by ionizing radiation (in Hungarian), *Kiserletugyi Kozl.* 63 E(1-3), 21, 1970.
164. **Paul, N., Grünewald, Th., and Kuprianoff, J.,** Über die Möglichkeit einer Behandlung von Trockensuppen mit ionisierenden Strahlen, *Dtsch. Lebensm. Rundsch.,* 65, 279, 1969.

165. **Nury, F. S., Miller, M. W., and Brekke, J. E.,** Preservative effect of some antimicrobial agents on high-moisture dried fruits, *Food Technol.,* 14, 113, 1960.
166. **Bolin, H. R., King, A. D., Jr., Stanley, W. L., and Jurd, L.,** Antimicrobial protection of moisturized Deglect Noor dates, *Appl. Microbiol.,* 23, 799, 1972.
167. **Pitt, J. I. and Christian, J. H. B.,** Water relation of xerophilic fungi isolated from prunes, *Appl. Microbiol.,* 16, 1853, 1968.
168. **Khan, I., Jan, M., Wahid, M., Neelofar, B., Atta, S., Akhtar, T., and Ahmad, A.,** Radiation preservation of dried fruits in Pakistan, abstract in *Food Irradiation for Developing Countries in Asia and the Pacific,* IAEA-TECDOC-271, International Atomic Energy Agency, Vienna, 1982, 136.
169. Irradiation of Plant Products, Comments from CAST, 1981-1, ISSN 0194-4096, Council for Agricultural Science and Technology, Budapest, April 1984.
170. **Emmerson, J. A., Kazanas, N., Gnaedinger, R. H., Krzeczkowski, R. A., Seagran, H. L., Markakis, P., Nicholas, R. C., Schweigert, B. S., and Kempe, L. L.,** Irradiation Preservation of Fresh-Water Fish and Inland Fruits and Vegetables, C00-1283-27, USAEC, Washington, D.C., 1965.
171. **Binstead, R. and Devey, J. D.,** *Soup Manufacture, Canning, Dehydration and Quick Freezing,* 3rd ed., Food Trade Press, London, 1970.
172. **Krugers-Dagneux, E. L. and Mossel, D. A. A.,** The microbiological condition of dried soups, in *Proc. 6th Int. Symp. Food Microbiol.,* Kampelmacher, E. H., Ingram, M., and Mossel, D. A. A., Eds., Grafische Industrie, Haarlem, The Netherlands, 1968, 411.
173. **Fanelli, M. J., Peterson, A. C., and Gunderson, M. F.,** Microbiology of dehydrated soups. I. A survey, *Food Technol.,* 19, 83, 1965.
174. **Nakamura, M. and Kelly, K. D.,** *Clostridium perfringens* in dehydrated soups and sauces, *J. Food Sci.,* 33, 424, 1968.
175. **Keoseyan, S. A.,** Incidence of *Clostridium perfringens* in dehydrated soup, gravy and spaghetti mixes, *JAOAC,* 54, 106, 1971.
176. **Powers, E. M., El-Bisi, H. M., and Rowley, D. B.,** Bacteriology of dehydrated space foods, *Appl. Microbiol.,* 22, 441, 1971.
177. Soup mix recalled because of *Salmonella, Food Chem. News,* p. 8, June 27, 1979.
178. Soup mix recalled because of *Salmonella* contamination, *Food Chem. News,* p. 52, October 14, 1974.
179. **Anema, P. J. and Michels, M. J. M.,** Microbiology of instant dry soup mixes (in Italian, English abstract), in *Proc. Int. Symp. Food Microbiol.,* Vol. 2, Federal Association of Science and Technology, Milan, 1974, 165.
180. **Mossel, D. A. A. and Krugers Dagneaux, E. L.,** Die hygienischbakteriologisch Beurteilung von Trockenkochsuppen, *Arch. Lebensm. Hyg.,* 14, 108, 1963.
181. **Coretti, K. and Müggenburg, H.,** Keimgehalt von Trockensuppen und seine Beureteilung, *Feinkostwirtschaft,* 4, 108, 1967.
182. **Gilbert, R. J. and Taylor, A. J.,** *Bacillus cereus* food poisoning, in *Microbiology in Agriculture, Fisheries and Food,* Skinner, F. A. and Carr, J. G., Eds., Academic Press, London, 1976, 197.
183. **Hobbs, W. E. and Greene, V. W.,** Cereal and cereal products, in *Compendium Methods for the Microbiological Examination of Foods,* Speck, M. L., Ed., American Public Health Association, Washington, D.C., 1976, 599.
184. International Commission on Microbiological Specifications for Foods, Cereals and Cereal Products, in *Microbial Ecology of Foods,* Vol. 2, Silliker, J. H., Elliott, R. P., Baird-Parker, A. C., Bryan, F. L., Christian, J. H. B., Clark, D. S., Olson, J. C., Jr., and Roberts, T. A., Eds., Academic Press, New York, 1980, 669
185. **Frazier, W. C.,** Contamination, preservation and spoilage of cereals and cereal products, in *Food Microbiology,* McGraw-Hill, New York, 1967, 180.
186. **Poisson, J., Jemmali, M., Cahagner, B., and Leclerc, J.,** The effect of gamma irradiation of wheat flour on its microflora and vitamin B_1 content, *Food Irradiat.,* 8, 2, 1967.
187. **Lorenz, K.,** Irradiation of cereal grains and cereal grain products (review), *Crit. Rev. Food Sci. Nutr.,* 6, 317, 1975.
188. **Zehnder, H. J. and Ettel, W.,** Zur Strahlenbehandlung von Lebensmittelzusatzstoffen und -halprodukten, *Alimenta,* 20, 67, 1981.
189. **Henon, Y.,** Past and recent events in food irradiation in France, *Food Irradiat. Newslett.,* 8(1), 41, 1984.
190. International Commission on Microbiological Specifications for Foods, Eggs, and Egg Products, in *Microbial Ecology of Foods,* Vol. 2, Silliker, J. H., Elliott, R. P., Baird-Parker, A. C., Bryan, F. L., Christian, J. H. B., Clark, D. S., Olson, J. C., Jr., and Roberts, T. A., Eds., Academic Press, New York, 1980, 521.
191. **Northolt, M. D., Wiegersman, N., and van Schotthorst, M.,** Pasteurization of dried egg white by high temperature storage, *J. Food Technol.,* 13, 25, 1978.

192. **Brogle, R. C., Nickerson, J. T. R., Proctor, B. E., Pyne, A., Campbell, C., Charm, S., and Lineweaver, H.,** Use of high voltage cathode rays to destroy bacteria of the *Salmonella* group in whole egg solids, egg yolk solids, and frozen egg yolk, *Food Res.*, 22, 572, 1957.
193. **Nickerson, J. T. R., Charm, S. E., Brogli, R. C., Lockhart, E. E., Proctor, B. E., and Lineweaver, H.,** Use of high-voltage cathode rays to destroy bacteria of the *Samonella* group in liquid and frozen egg white and egg white solids, *Food Technol.*, 11, 159, 1957.
194. **Thornley, M. J.,** Microbiological aspects of the use of radiation for the elimination of salmonellae from foods and feeding stuffs, in *Radiation Control of Salmonella in Food and Feed Products,* Tech. Rep. Ser. No. 22, International Atomic Energy Agency, Vienna, 1963, 81.
195. **Kahan, R. S.,** Experimental irradiation of dehydrated foodstuffs and animal feeds in Israel, in *Microbiology of Dried Foods,* Kampelmacher, E. H., Ingram, M., and Mossel, D. A. A., Eds., International Association of Microbiological Societies, New York, 1969, 301.
196. **Inghilesi, E., Piccininno, G., Tiecco, G., and Cacciapuoli, B.,** The use of gamma radiations for the radicidation of dried egg albumen (in Italian), *Riv. Sci. Tecn. Alim. Um.*, 5, 299, 1975.
197. **Bomar, M. T.,** Salmonellenbekämpfung in Trockenei durch Bestrahlung, *Arch. Lebensmittelhyg.*, 21(5), 97, 1970.
198. **Diehl, J. F.,** Elektronenspinresonanz — Untersuchungen an strahlenkonservierten Lebensmitteln. II. Einfluss des Wassergehaltes auf die Spinkonzentration, *Lebensm. Wiss. Technol.*, 5, 51, 1972.
199. **Katušin-Ražem, B.,** Possibilities of Eradication of Salmonellae in Whole Egg Powder by Ionizing Radiation, Prog. Rep. December 1983—August 1984, IAEA Res. Contract No. 3636/RB, "Ruder Bosković" Institute, Zagreb, Yugoslavia, 1984.
200. **Diehl, J. F.,** Verminderung von Strahleninduzierten Vitamin-E und B_1-Verlusten durch Bestrahlung von Lebensmitteln bei tiefen Temperaturen und durch Ausschuss von Luftsauerstoff, *Z. Lebensm. Unters. Forsch.*, 169, 276, 1979.
201. **Galesloot, T. E. and Stadhouders, J.,** The microbiology of spray dried milk products with special reference to *Staphylococcus aureus* and salmonellae, in *Microbiology of Dried Foods,* Kampelmacher, E. H., Ingram, M., and Mossel, D. A. A., Eds., International Association of Microbiological Societies, New York, 1969, 313.
202. **Lucaz, M.,** The effect of gamma (^{60}Co) rays upon the changes of fat in the whole milk powder, *Rocz. Inst. Przem. Mlecz.*, 14(3), 49, 1972.
203. International Commission on Microbiological Specifications for Foods, Sugar, Cocoa, Chocolate, and Confectioneries, in *Microbial Ecology of Foods,* Vol. 2, Silliker, J. H. B., Elliott, R. P., Baird-Parker, A. C., Bryan, F. L., Christian, J. H. B., Clark, D. S., Olson, J. C., Jr., and Roberts, T. A., Eds., Academic Press, New York, 1980, 778.
204. **Mossel, D. A. A., Meursing, E. H., and Slot, H.,** An investigation on the numbers and types of aerobic spores in cocoa powder and whole milk, *Neth. Milk Dairy J.*, 28, 149, 1974.
205. **Gabis, D. A., Langlois, B. E., and Rudnick, W.,** Microbiological examination of cocoa powder, *Appl. Microbiol.*, 20, 644, 1970.
206. **D'Aoust, J. Y.,** *Salmonella* and the chocolate industry. A review, *J. Food Prot.*, 40, 718, 1977.
207. WHO, *Wkly. Epidemiol. Rec. World Health Organ. Geneva,* 48, 377, 1973.
208. **Gruünewald, Th. and Münzer, R.,** Strahlenbehandlung von Kakaopulver, *Lebensm. Wiss. Technol.*, 5(6), 203, 1972.
209. **Schaffner, C. P., Mosbach, K., Bibit, V. C., and Watson, C. H.,** Coconut and *Salmonella* infection, *Appl. Microbiol.*, 15, 471, 1967.
210. **Ley, F. J., Freeman, B. J., and Hobbs, B. C.,** The use of gamma radiation for the elimination of salmonellae from various foods, *J. Hyg. Camb.*, 61, 515, 1963.
211. **Jacobs, G. P. and Eisenberg, E.,** The reconstitution of powders for injection with γ-irradiated water, *Int. J. Appl. Radiat. Isot.*, 32, 180, 1981.
212. **Rogachov, V. I., Kardashev, A. V., Osipov, V. B., Bregvadze, V. D., Kovalskaya, L. P., Malischev, S. D., Korotchenko, K. A., Pootinova, M. N., Frumkin, N. L., Dobrovskaja, N. D., Golovkova, G. N., Govtarn, I. M., Boockteyev, V. A., Vikulin, A. A., and Makarova, L. P.,** The use of ionizing radiation in food industry, A/CONF.49/P/696, in *4th U.N. Int. Conf. on Peaceful Uses of Atomic Energy,* Vol. 12, United Nations, New York, 1972.
213. **Kokubo, Y., Jimbo, K., Murakami, F., and Murakami, H.,** Prevalence of spore-forming bacteria in sugar, starch, spices and vegetable proteins, *Ann. Rep. Tokyo Metr. Res. Lab. P.H.*, 33, 155, 1982.
214. **Saint-Lébe, L. and Berger, G.,** Preservation of powdered food products by irradiation: advantages and possibilities of the method: application to the case of maize starch, A/CONF.49/P/628, in *4th U.N. Int. Conf. on Peaceful Uses of Atomic Energy,* Vol. 12, United Nations, New York, 1972.
215. **Fretton, R., Delattre, J. M., and Beerens, H.,** Irradiation gamma d' échantillons d'amidon contamines par des sporulés anaérobies, *Starke,* 29(1), 5, 1971.

216. **Saint-Lèbe, L., Mucchielli, A., Leroy, P., and Beerens, H.,** Etudes preliminaires de la microflore de l'amidon de mais avant et apres irradiation, in *Radiation Preservation of Food,* International Atomic Energy Agency, Vienna, 1973, 155.
217. **Saint-Lébe, L., Berger, G., and Muchielli, A.,** Influence d'une irradiation gamma sur la salubrite et les proprietes technologiques de l'amidon de mais, in *Improvement of Food Quality by Irradiation,* International Atomic Energy Agency, Vienna, 1974, 51.
218. **Fretton, R., Fretton, J., Delattre, J. M., and Beerens, H.,** Evolution de la microflore de deux echantillons d'amidon apres irradiation gamma, *Starke,* 27(1), 4, 1975.
219. **Delattre, J. M., Fretton, J. M., Fretton, R., Poncelet, F., and Beerens, H.,** Etude de l'irradiation gamma de poudres alimentaires, effects sur la microflore aérobic de l'amidon, *Eur. J. Appl. Microbiol.,* 1, 259, 1975.
220. **Berger, G., Agnel, J. P., and Saint Lébe, L.,** Etude de la partie soluble dans l'eau des radiodextrines formees par irradiation gamma de l'amidon de mais, *Starke,* 29(2), 40, 1977.
221. **Korotchenko, K. A., Putilova, I. N., Akulova, I. S., and Chernobaeva, N. N.,** The possible utilization of products from radiolysis of starch (in Russian), in *Radiatsionnaya Obrabnotka Pishchevykh Produktov,* Collection of Reports, Gosudarstvennyj Komitet pro Ispol'zovaniyu Atomnoj Energii, Moscow, 1971, 169.
222. **Skinner, E. R. and Kertesz, Z. I.,** The effect of gamma radiation on the structure of pectin. An electrophoretic study, *J. Polym. Sci.,* 47, 99, 1960.
223. **Dimitrova, N.,** Decontamination of gelatine by ionizing radiation. I. Physico-chemical properties and organoleptic changes of gelatine irradiated at various gamma radiation doses (in Bulgarian), *Mesopromishlenost Bull.,* 6 (1—2), 31, 1973; as cited in *Z. Lebensm. Unters. Forsch.,* 156(4), 252, 1974.
224. **Dimitrova, N.,** Decontamination of gelatine by ionizing radiation. II. Microbiological changes in gelatine irradiated at various gamma radiation doses (in Bulgarian), *Mesopromishlennost Bull.,* 6(3—4), 10, 1973, as cited in *Z. Lebensm. Unters. Forsch.,* 156(4), 252, 1974.
225. **Bachman, S., Galant, S., Gasyna, Z., Witkowski, S., and Zegota, H.,** Effects of ionizing radiation on gelatin in the solid state, in *Improvement of Food Quality by Irradiation,* International Atomic Energy Agency, Vienna, 1974, 77.
226. **Frank, H. K. and Grünewald, Th.,** Untersuchungen über die Möglichkeiten einer Strahlensterilisierung (Radappertization) von Gelatine, *Fleischwiurtschaft,* 49(1), 74, 1969.
227. **Mateles, R. L. and Goldblith, S. A.,** Some effects of ionizing radiations on gelatin, *Food Technol.,* 12, 633, 1958.
228. **Brankova, R. and Dimitrova, N.,** Reduction of bacterial contamination of protein preparations by ionizing radiation (in Bulgarian), *Khran. Prom.,* 24(7), 21, 1975.
229. **Dimitrova, R. and Brankova, N.,** Reduction of bacterial contamination of protein preparations by ionizing radiation (in Bulgarian), *Khran. Prom.,* 24(6), 21, 1975.
230. La Debacterisation de Sang, Plasma, et Cruor Deshydrates par Traitement au Rayonnement Gamma, Dossier de Demande d'Autorisation, Institute National des Radioelements-IRE, Fleurus, 1984.
231. **Tsuji, K.,** Low-dose cobalt-60 irradiation for reduction of microbial contamination in raw materials for animal health products, *Food Technol.,* 37(2), 48, 1983.
232. **Kerekes, L.,** Trend of the microbiological quality of sugar (in Hungarian), *Élelmiszervizsgálati Közl.,* 30, 53, 1984.
233. **Scarr, M. P.,** Symposium on growth of microorganisms at extremes of temperature, thermophiles in sugar, *J. Appl. Bacteriol.,* 31, 66, 1968.
234. **Sabine, F. M.,** Application of Ionizing Radiation to Destroy the Bacteriological Contamination of Sugar and Related Carbohydrates, Prog. Rep. U.S. Army Quartermaster Com., No. 34, American Crystal Sugar Company, 1956.
235. **Kiss, I., Farkas, J., Andrássy, É., and Beczássy, K.,** Observations on radiation sterilized sugar and its microbiological effects, *Acta Microbiol. Pol.,* 17, 67, 1968.
236. **Ravishankar, D. and Bhave, R. N.,** A comparative study on aqualuminescence, chemical effects and photoannealing of γ-irradiated NaCl and K_2SO_4, *J. Radioanal. Nucl. Chem. Lett.* 87, 331, 1984.
237. **Turjanski, C., Kramar, E., Mugliaroli, Y. H. A., Pezzolanti, R., Casos, C. C., and Ambesi, A.,** Irradiacion de albumine de sangre desecada de bovino: estudio bacteriologico, in *Food Irradiation,* International Atomic Energy Agency, Vienna, 1966, 321.
238. **Lott, G. and Frank, H. K.,** Mikrobiologische Verunreinigungen in Enzympräparaten für die Lebensmitteltechnologie, *Dtsch. Lebensm. Rundsch.,* 69, 73, 1973.
239. **Vas, K.,** Effects of ionizing radiations on pectic enzymes, in *Proc. 1st In. Congr. Food Sci. Technol. London,* Vol. 1, Gordon & Breach, New York, 1962, 625.
240. **Vas, K.,** Effects of ionizing radiations on mould cells and on their pectic enzymes (in Hungarian), *Kert. Szolesz. Foiskola Kozl.,* 2, 307, 1964.
241. **Vas, K.,** A study of the relative radiosensitivities of moulds and their pectic enzymes, in *IAEA Research Contracts, Fifth Annual Report,* IAEA Tech. Rep. Ser. No. 28, International Atomic Energy Agency, Vienna, 1964, 53.

242. **Vas, K. and Proszt, G.**, Mikrobiologische Stabilisierung von Pektin Enzympräparaten durch ionisierende Strahlen, *Ber. Wiss. Tech. Komm. Int. Fruchtsaft. Union (Luzerner Ber.)*, 6, 449, 1965.
243. **Delincée, H., Münzner, R., and Radola, B. J.**, Bestrahlung von technischen Enzympräparaten für die Lebensmitteltechnologie, *Lebensm. Wiss. Technol.*, 8, 270, 1975.
244. **Diehl, F. J.**, *Bestrahlung von Enzympräparaten*, Informationsheft No. 101 des Büros EURISOTOP, Information und Dokumentation des Büros EURISOTOP, Brussels, 1975.
245. **Leuchtenberger, A. and Rutloff, H.**, Zur Keimminderung von technischen Enzympräparaten unter besonderer Beräparaten unter besonderer Berücksichtigung der Behandlung mit Gamma Strahlen, *Nahrung*, 20, 525, 1976.
246. **Kawashima, K., Nango, Y., and Umeda, K.**, Irradiation of enzymes. I. Radiosterilization of amylase, *Food Irradiat. Jpn.*, 7(1), 84, 1972.
247. **Hespeels, L.**, Progress in food irradiation — Belgium, *Food Irradiat. Inf.*, 12, 13, 1982.
248. **Diehl, J. F.**, Progress in food irradiation — Germany, *Food Irradiat. Inf.*, 12, 26, 1982.
249. **Ribarić, J., Briski, B., Matić, S., Ražem, D., Katalenić, M., Mihoković, N., and Beljak, K. J.**, Preservation of commercial enzyme preparations by irradiation, presented at the ESNA Working Group on Food Irradiation, Piacenza, Italy, September 3 to 7, 1984.
250. **Pandula, E. L., Farkas, E., and Nagykáldi, A.**, Final Report of the Agency, Res. Contract No. 351/RB, Pharmaceutical Institute of the Medical University, Budapest, 1968.
251. **Craeghs, M.**, Radiological Inactivation of Papain by Gamma Radiation (in Flemish), Thesis, Katholieke Universiteit te Leuven, Belgium, 1980.
252. **Patel, K. M. and Gopal, N. G. E.**, Effect of gamma radiation and ethylene oxide on papain, *Indian J. Pharm. Sci.*, 41(2), 81, 1979.
253. **Samoilenko, I. I., Fedotov, N. S., Tumanyan, M. A., and Korolev, N. I.**, Use of combined radiation methods for decreasing the bacterial count of enzymatic preparations (in Russian), *Prik. Biokhim. Mikrobiol.*, 20(2), 239, 1984.
254. **Bachman, St. and Borkowska, M.**, Changes in the enzymatic activity of rennin in different purity under the influence of ^{60}Co gamma irradiation, *Rocz. Technol. Chem. Zywnosci.*, 22, 123, 1972; as cited in *Nucl. Sci. Abstr.*, 29, 26969, 1974.
255. **Zetelaki-Horváth, K. and Kiss, I.**, Radiation effects on activity and storage stability of endo-poygalacturonase, *Acta Ailment.*, 7, 299, 1978.
256. **Quehl, A., Leuchtenberger, A., and Schalinatus, E.**, Entkeimung von Pankreatinpräparaten mittels Gamma-Strahlen, *Nahrung*, 29, 105, 1985.
257. **Bachman, S. and Gebicka, L.**, Effect of gamma irradiation on whole-cell glucose isomerase. I. Gamma-rays induced inactivation of whole-cell glucose isomerase, *Starch/Starke*, 36(3), 94, 1984.
258. **Bachman, S. and Gebicka, L.**, Effect of gamma irradiation on whole-cell glucose isomerase. II. Properties of irradiated whole-cell glucose isomerase, *Starch/Starke*, 36(6), 212, 1984.
259. **Vas, K. and Proszt, G.**, Radiosensitivity of purified and crude pectic enzyme preparations, in *Application of Food Irradiation in Developing Countries*, IAEA Tech. Rep. Ser. No. 54, International Atomic Energy Agency, Vienna, 1966, 151.
260. **Farkas, J.**, Response of bacterial spores to combination of heat, irradiation and some chemical factors controlling the spore state, presented at the Food and Agric. Organ./Int. At. Energy Agency Res. Coordination Meet. on Microbiol. Aspects of Food Irradiat., International Atomic Energy Agency, Vienna, 1970.
261. **Farkas, J., Beczner, J., Kiss, I., and Incze, K.**, Cell Count Reduction in Seasoning, Particularly in Ground Paprika, by Radiation Treatment, 3rd Prog. Rep. to International Atomic Energy Agency, Contract No. 931/R1/RB, January 1, 1972 to June 30, 1972, Central Food Research Institute, Budapest, 1972.
262. **Farkas, J. and Andrássy, É.**, Decrease of bacterial spoilage of bread by low-dose irradiation of its flour, in *Combination Processes in Food Irradiation*, International Atomic Energy Agency, Vienna, 1981, 81.
263. **Farkas, J. and Andrássy, É.**, Increased sensitivity of surviving bacterial spores in irradiated spices, in *Fundamental and Applied Aspects of Bacterial Spores*, Dring, G. J., Ellar, D. J., and Gould, G. W., Eds., Academic Press, London, 1985, 397.
264. **Morgan, B. H. and Reed, J. M.**, Resistance of bacterial spores to gamma irradiation, *Food Res.*, 19, 357, 1954.
265. **Roberts, T. A., Ditchett, M., and Ingram, M.**, The effect of sodium chloride on radiation resistance and recovery of irradiated anaerobic spores, *J. Appl. Bacteriol.*, 28, 336, 1965.
266. **Roberts, T. A.**, Symposium on bacterial spores: recovering spores damaged by heat, ionizing radiations and ethylene oxide, *J. Appl. Bacteriol.*, 33, 74, 1970.
267. **van Zuilichem, D. J., Stolp, W., and Beumer, R. R.**, Application of infrared-radiation in food industries, in Collected Descriptions of the Contributions at the Exhibitions Food Engineering '83 — Process Equipment '83, Utrecht, Holland, November 21 to 25, 1983, 39.

268. **Szabad, J. and Kiss, I.**, Comparative studies on the sanitizing effects of ethylene oxide and of gamma radiation in ground paprika, *Acta Aliment.*, 8, 383, 1979.
269. **Tipples, K. H. and Norris, F. W.**, Some effects of high level gamma irradiation on the lipids of wheat, *Cereal Chem.*, 42, 437, 1965.
270. **Borris, R. and Woelbing, M.**, Fachdiscussion zur toxikologischer Untersuchung bestrahlter Enzympräparate, *Lebensm. Ind.*, 27, 524, 1980.
271. **Farkas, J., Török, G., and Horváth, Zs.**, Storage experiments with ground paprika pasteurized by ionizing radiation, (in Hungarian), *Commun. Cent. Food Res. Inst. Budapest,* (2), 19, 1962.
272. **Farkas, J.**, Radurization and radiation of spices, in *Aspects of the Introduction of Food Irradiation in Developing Countries,* International Atomic Energy Agency, Vienna, 1973, 43.
273. **Brower, J. H. and Tilton, E. W.**, Insect disinfestation of dried fruit by using gamma radiation, *Food Irradiat.*, 11(1—2), 10, 1970.
274. **Beczner, J. and Farkas, J.**, Investigations into the radioresistance of Plodia interpunctella (Hübner), *Acta Phytopathol. Acad. Sci. Hung.*, 9 (1—2), 153, 1974.
275. **Tilton, E. W. and Burditt, A. K.**, Insect disinfestation of grain and fruit, in *Preservation of Food by Ionizing Radiation,* Vol. 3, Josephson, E. S. and Peterson, M. S., Eds., CRC Press, Boca Raton, Fla., 1983, 215.
276. **Dutova, W. N., Kardashev, A. V., and Goftarsch, M. M.**, The use of gamma irradiation in cold sterilization of salt and spice mixtures, *All-Union Res. Mar. Fish. Oceanogr. Moscow,* 73, 69, 1970 (English translation, Fisheries Research Board, Canada).
277. **Weiss, D., Weiss, B., Doelstadt, R., and Huebner, G.**, A comparative study of the production of preserves using irradiated spice mixtures (7.5 and 10 kGy) and commercially available spices, in *ZFI-Mitteilungen,* No. 98, Zentralinstitut für Isotopen- und Strahlenforschung, Leipzig, 1984, 648.
278. **Screenivasan, A.**, Compositional and quality changes in some irradiated foods, in *Improvement of Food Quality by Irradiation,* International Atomic Energy Agency, Vienna, 1974, 129.
279. **Masek, J., Miková, L., Teply, M., Havlová, J., and Hošek, K.**, Effects of gamma irradiation upon the proteases of rennets and their surrogates (in Czech), *Prumysl. Potravin,* 22, 220, 1971.
280. **Scholze, U. and Grünewald, Th.**, Qualitätsbeurteilung von bestrahlten Trockenhülsenfrüchten, und Graupen, *Ind. Obst- GemüseVerwert.*, 53, 215, 1968.
281. **Kohn, R. M.**, Produktverbesserungen an ausgewählten entwässerten Lebensmitteln und sonstigen Naturerzeugnissen durch Einwirkung ionisierender Strahlen, *Lebensm. Wiss. Technol.*, 43, 69, 1971.
282. **Rao, V. S. and Vakil, U. K.**, Effects of gamma-radiation on cooking quality and sensory attributes of four legumes, *J. Food Sci.*, 50, 372, 1985.
283. **Penner, H.**, Der gegenwärtige Stand der Forschung und Technik auf dem Gebiet der Strahlenkonservierung pflanzlicher Lebensmittel, *Ind. Obst-Gemüseverwert.*, 52, 321, 1967.
284. **Karuo, C., Katsuichi, K., and Kuniyasu, O.**, Effect of gamma radiation on the flavour of horticultural products — effects on apple pulp, apple jam and dried bananas (in Japanese) *Food Irradiat. Jpn.*, 4, 1, 77, 1969.
285. **Kempe, L. L., Graikoski, J. T., Stratton, J. R., and Day, W. H.**, Sterilization of barley malt with gamma radiation, *Agric. Food Chem.*, 12, 98, 1964.
286. **Hsu, H., Hadziyev, D., and Wood, F. W.**, Gamma irradiation effect on the sulfhydryls content of skim milk powder, *Can. Inst. Food Sci. Technol. J.*, 5, 191, 1972.
287. **Hsu, H., Hadziyev, D., and Wood, F. W.**, Gamma irradiation effect on some volatiles of skim milk powder, *Can Inst. Food Sci. Technol. J.*, 5, 197, 1972.
288. **Samec, M.**, Der Einfluss von Gammastrahlen und Ultraschall auf Stärke, *Starke,* 12, 165, 1960.
289. **Samec, H.**, Vergleichender Abbau verschiedener Stärken durch Gammastrahlen, Elektronenstrahlen und Ultraschall, *Starke,* 13, 283, 1961.
290. **Hofreiter, B. T. and Russell, C. R.**, Gamma-irradiated corn starches, alkaline dispersions for surface-sizing paper, *Starke,* 26, 18, 1974.
291. **Džamic, M. D. and Jankovic, B. R.**, Radiation effects in pectins, *Int. J. Appl. Radiat. Isot.*, 17, 561, 1966.
292. **Fritsch, G. and Reymond, D.**, Effects of X-rays on pectin studies by electron spin resonance, *Int. J. Appl. Radit. Isot.*, 21, 329, 1970.
293. The Use of Gamma Radiation Sources for the Sterilization of Pharmaceutical Products, Reports of a Working Party, Association of British Pharmaceutical Industry, London, June 1960.
294. **Artandi, Ch. and Van Winkle, W., Jr.**, Sterilization of pharmaceuticals and hospital supplies by ionizing radiation, in *Large Radiation Sources in Industry,* International Atomic Energy Agency, 1960, 249.
295. Radiation Sterilization of Pharmaceuticals (In Hungarian), State Commission for Technical Development, 6-7405-Et, Országos Müszaki Fejlesztési Bizottság, Budapest, July 1975.

296. **Kawashima, K., Tanaka, Y., and Umeda, J.,** Irradiation of enzyme preparations, II. Radiosterilization of enzyme preparations by accelerated electron irradiation (in Japanese), Report of the National Food Research Institute (Shokuryo Kenkyusho Kenkyu Hokoku) No. 30, 83, 1975; *Food Sci. Technol. Arch.*, 8 (1), 1976.
297. **Kawashima, K., Nango, Y., Doi, T., and Umeda, K.,** *J. Food Sci. Technol. (Tokyo),* 20, 9, 1973, as cited in **Diehl, F. J.,** Bestrahlung von Enzympräparaten, Informationsheft No. 101 des Büros EURISOTOP, Information und Dokumentation des Büros EURISOTOP, Brussels, 1975.
298. **Kawashima, K., Tanaka, Y., and Umeda, K.,** *J. Food Sci. Technol. (Tokyo),* 21, 592, 1974; as cited in **Diehl, F. J.,** Bestrahlung von Enzympräparaten, Informationsheft No. 101 des Büros EURISOTOP, Information und Dokumentation des Büros EURISOTOP, Brussels, 1975.
299. **Sanner, T., Kovács-Proszt, G., and Witkowsi, S.,** Aspects of the effect of ionizing radiation on enzymes, in *Improvement of Food Quality by Irradiation,* International Atomic Energy Agency, Vienna, 1974, 101.
300. **Pihl, A. and Sanner, T.,** X-ray inactivation of papain in solution, *Radiat. Res.*, 19, 27, 1963.
301. **Kovács-Proszt, G. and Sanner, T.,** X-ray inactivation in solution of rennin from *Mucor pusillus, Radiat. Res.*, 53, 444, 1973.
302. **Delincée, H., Radola, B. J., and Drawert, F.,** The effect of combined heat and irradiation treatment on the isoelectric and size properties of horse-radish peroxidase, *Acta Aliment.*, 2, 259, 1973.
303. **Kawashima, K. and Umeda, K.,** Immobilization of enzymes by the radiopolymerization of acryl amide, in *Improvement of Food Quality by Irradiation,* International Atomic Energy Agency, Vienna, 1974, 119.
304. **Ahmed, M. S. H., Hameed, A. A., Kadhum, A. A., Ali, S. R., Farkas, J., Langerak, D. Is., and Van Duren, M. D. A.,** Comparative evaluation of trial shipments of fumigated and radiation disinfested dates from Iraq, *Acta Aliment.*, 14, 355, 1985.
305. **Kwon, J. H., Byun, M. W., and Cho, H. O.,** Sterilization of garlic powder by irradiation (in Korean), *Korean J. Food Sci. Technol.*, 16, 139, 1984, as cited in *Food Sci. Technol. Arch.*, 17(5), 253, abstr. no. 5T13.
306. Irradiation proposal seen as "pragmatic and logical" by spice group, *Food Chem. News,* p. 31, May 7, 1974.
307. **Toofanian, F., Stegeman, H., and Streutjens, E.,** Comparative effect of ethylene oxide and gamma irradiation on the chemical, sensory and microbial quality of ginger, cinnamon, fennel and fenugreek, IFFIT Rep. No. 60, International Facility for Food Irradiation Technology, Wageningen, August 1985.
308. **de Boer, E. and Boot, E. M.,** Comparison of methods for isolation and confirmation of *Clostridium perfringens* from spices and herbs, *J. Food Prot.*, 46, 533, 1983.
309. **Flannigan, B. and Hui, S. C.,** The occurrence of aflatoxin-producing strains of *Aspergillus flavus* in the mould floras of ground spices, *J. Appl. Bacteriol.*, 41, 411, 1976.
310. **Padwal-Desai, S. R., Munasiri, M. A., Ghankar, A. S., Sharma, A., and Nadharni, N. R.,** The improvement of quality of spices by gamma irradiation, Poster IAEA-SM-271/23P, prepared for the Int. Symp. on Food Irradiat. Processing, Washington, D.C., March 4 to 8, 1985.
311. **Morre, I., Serres, L., and Janin, F.,** Emploi des radiations ionisantes en technologie laitière, *Lait,* 58, 381, 1978.
312. The Effect of Irradiation upon Spices. II. The Effect of Radiation Treatment on the Lipid Composition of Several Spices, 1st Interim Rep. by the Central Food Research Institute, Budapest, and the Department of Agricultural Chemical Technology, Technical University of Budapest, Hungary, IFIP-R 47, International Project in the Field of Food Irradiation, Karlsruhe, West Germany, April 1978.
313. The Effect of Ionizing Radiation on the Carotenoid Content of Paprika, Rep. to the International Project in the Field of Food Irradiation by the Central Food Research Institute and the Isotope Institute of the Hungarian Academy of Sciences, Budapest, May 1980.
314. **Ismail, F. A.,** Effect of irradiation on broad bean *(Vicia faba)* textural qualities, *Lebensm. Wiss. Technol.*, 9, 18, 1976.
315. Annual Report 1981, Danish Meat Research Institute, Roskilde, Denmark, 1982, 31.
316. **Kispéter, J., Cséfalvay, M., and Fenyvessy, J.,** Effect of gamma radiation on the characteristics of milk powder, in *Utilization of Radiation Energy in the Food Industry and in Agriculture,* Central Food Research Institute, Buadpest, 1983, 23.
317. **Rosenthal, I., Martinot, M., Lindner, P., Juven, B. L., and Ben-Hur, E.,** A study of ionizing irradiation of dairy products, *Milchwissenschaft,* 38, 467, 1983.
318. **Bockemühl, J. and Wohlers, B.,** Zur Problematik der Kontamination unbehandelter Trockenprodukte der Lebensmittelindustrie mit Salmonellen, *Zentralbl. Bakteriol. Parasitenk. Infektionskr. Hyg. Abt. 1 Orig. Reihe B,* 178, 535, 1984.
319. Centers for Disease Control, Outbreak of *Salmonella oranienburg* infection — Norway, *Morb. Mort. Wkly. Rep.*, 31, 655, 1982.
320. **Farkas, J. and Andrássy, É.,** Comparative effects of gamma irradiation and ethylene oxide treatments on the quality of selected spices, presented at the 16th Annu. Meet. of the Eur. Soc. of Nucl. Methods in Agric., Warsaw, September 9 to 13, 1985.

321. **Morrison, R. M. and Roberts, T.,** *Food Irradiation: New Perspectives on a Controversial Technology. A Review of Technical, Public Health, and Economic Considerations,* Office of Technology Assessment, Congress of the U.S. Washington, D.C., December 1985.
322. **Patel, K. M., Tantry, M., Sharma, G., and Gopal, N. G. S.,** *Indian J. Pharm. Sci.,* 41, 209, 1979; as cited in **Jacobs, G. P.,** *Radiat. Phys. Chem.,* 26, 133, 1985.
323. **Jacobs, G. P.,** A review: radiation sterilization of pharmaceuticals, *Radiat. Phys. Chem.,* 26, 133, 1985.
324. **Jacobs, G. P. and Simes, R.,** *J. Pharm. Pharmacol.,* 31, 333, 1979; as cited in **Jacobs, G. P.,** *Radiat. Phys. Chem.,* 26, 133, 1985.
325. **Achmatowicz-Szmajke, T., Bryl-Sandelewska, T., and Galazka, M.,** *Radiochem. Radional. Lett.,* 38, 5, 1979, as cited in **Jacobs, G. P.,** *Radiat. Phys. Chem.,* 26, 133, 1985.
326. **Chang, B.-L., Haney, W. G., and Nuessle, N. O.,** *J. Pharm. Sci.,* 63, 758, 1974; as cited in **Jacobs, G. P.,** *Radiat. Phys. Chem.,* 26, 133, 1985.
327. **Prince H. N. and Welt, M. A.,** *Am. Perfum. Cosmet.,* 86, 49, 1971; as cited in **Jacobs, G. P.,** *Radiat. Phys. Chem.,* 26, 133, 1985.
328. **Niemand, J. G.,** Food-borne pathogens: is there a remedy?, *Food Ind. South Afr.,* March 1985.
329. **Bachman, S.,** *Changes in Activity of Industrial Enzyme Preparations Irradiated with Sterilizing Doses,* Part of a coordinated programme on factors influencing the utilization of food irradiation process, Final rep. for the period May 1, 1982 to April 30, 1984, IAEA-R-3097-F, International Atomic Energy Agency Vienna, March 1984.
330. **Hartgen, H. and Kahlan, D. R.,** Bedeutung der Koloniezahl bei Haushaltsgewürzen, *Fleischwirtschaft,* 65(1), 99, 1985.
331. **Muhamed Lebai Juri, Ito, H., Watanabe, H., and Tamura, N.,** Distribution of microorganisms in spices and their decontamination by gamma-irradiation, *Agric. Biol. Chem.,* 50, 347, 1986.
332. **Kovacs, N.,** Salmonellae in desiccated coconut, egg pulp, fertilizer, meat-meal, and mesenteric glands: preliminary report, *Med. J. Aust.,* 46, 557, 1959.
333. **Gustafsen, S. and Breen, O.,** Investigation of an outbreak of *Salmonella oranienburg* infection in Norway caused by contaminated black pepper, *Am. J. Epidemiol.,* 806, 1984.
334. **Bullerman, L. B., Lieu, F. Y., and Seier, S. A.,** Inhibition of growth and aflatoxin production by cinnamon and clove oils, cinnamic aldehyde and eugenol, *J. Food Sci.,* 42, 1107, 1977.
335. **Morozumi, S.,** Isolation, purification and antibiotic activity of *o*-methoxycinnamaldehyde from cinnamon, *Appl. Environ. Microbiol.,* 36, 577, 1978.
336. **Brower, J. H. and Tilton, E. W.,** Insect disinfestation of dried fruit by using gamma radiation, *Food Irradiat.,* 11(1—2), 10, 1970.
337. **Saravacos, G. and Macris, B.,** Radiation preservation of grapes and some other Greek fruits, *Food Irradiat.,* 4(1—2), A19, 1963.
338. **Jacquet, J. and Teherani, M.,** An unusual presence of aflatoxin in certain animal products. Possible role of pepper, *Bull. Acad. Vet. Fr.,* 47, 313, 1974.
339. **Hitokoto, H., Morozumi, S., Wauke, T., Sakai, S., and Kurata, H.,** Fungal contamination and mycotoxin detection of powdered herbal drugs, *Appl. Environ. Microbiol.,* 36, 252, 1978.
340. **Clampet, G. L., Ed.,** *Agricultrual Statistics 1983,* U.S. Department of Agriculture, Washington, D.C., 1983.
341. **Heath, H. B.,** *Source Book of Flavours,* AVI Publishing, Wesptort, Conn., 1981, 99.
342. **Fenwick, G. R. and Hanley, A. B.,** The genus allium. I., *Crit. Rev. Food Sci. Nutr.,* 22(3), 199, 1985.

Chapter 3

SOME TECHNOLOGICAL AND ENGINEERING ASPECTS OF RADIATION DECONTAMINATION

I. INTRODUCTION

Radiation processing is a relatively simple technology. Materials to be treated are exposed to a controlled amount of radiation (ionizing energy) from a radiation source. Irradiation facilities can be operated as a part of the food processing lines or handling lines of storehouses, or they can be operated as service facilities on a fee basis. The necessary size of the facility is governed by the dose requirement for the product and the number of kilograms put through the facility per hour plus the number of hours per year it is to be operated

II. RADIATION SOURCES

Gamma ray-emitting isotopic sources, i.e., ^{60}Co and ^{137}Cs, and machine sources, i.e., certain electron accelerators and X-ray devices, can be used as radiation sources for food processing applications. Ionizing radiations produced by these radiation sources have the same biological effects.

^{60}Co is a man-made isotope produced from the stable metal ^{59}Co in specific nuclear reactors.[1,137] Cs is a fission product, and it is a component of nuclear waste resulting from power generation or production of plutonium. Producing ^{137}Cs in usable forms requires reprocessing of spent nuclear fuel.

^{60}Co is a more energetic producer of γ-rays than ^{137}Cs. The energy of gamma photons from ^{60}Co is 1.17 and 1.33 MeV, while the gamma energy of ^{137}Cs is 0.66 MeV.

^{60}Co has a half-life of 5.3 years, while the half-life on ^{137}Cs is 30.2 years, so the latter requires much less replenishment than ^{60}Co to maintain a constant throughput capacity at the irradiation facility in the course of years of its use. ^{60}Co plants require an addition of about 12% of the original source activity every year, while ^{137}Cs maintains a reasonably constant flux field if 12% of the original activity is added every 5 years. On the other hand, due to their different physical characteristics (metallic cobalt or cobalt oxide vs. cesium chloride as source material), ^{137}Cs sources require a more sophisticated control system. For use, the radioisotopes are doubly encapsulated in stainless steel units, which are grouped together to form the actual source of radiation of an irradiation facility.

When a beam of high-energy electrons collides with a dense metal plate (i.e., tungsten or gold), X-ray (Bremsstrahlung) is produced within the metal plate when the electrons are stopped.[2] γ- and X-rays are the same type of electromagnetic radiations. There are, however, differences between Bremsstrahlung and γ-radiation which influence their utility: X-ray sources produce the electromagnetic radiation in a broad spectrum, while γ-rays from isotopic sources are monoenergetic.

γ-Rays are emitted in all directions, unlike the high-energy Bremsstrahlung, which is emitted preferentially in the direction of the electron beam which produces it (forward scattering property).

γ- and X-rays have a high penetrating capacity. Compared to them, accelerated electrons have a much more limited penetration capacity. The penetration of high-speed electrons depends on the voltage (the energy of the electrons) and, of course, on the density (ρ) of the product, and at bulk densities of dry food ingredients only relatively thin layers (no more than several centimeters) can be treated. Figure 8 compares the penetration of a 3-MeV electron beam and ^{60}Co γ-irradiation, respectively, in flour with a bulk density of 0.74 g/cm.[33]

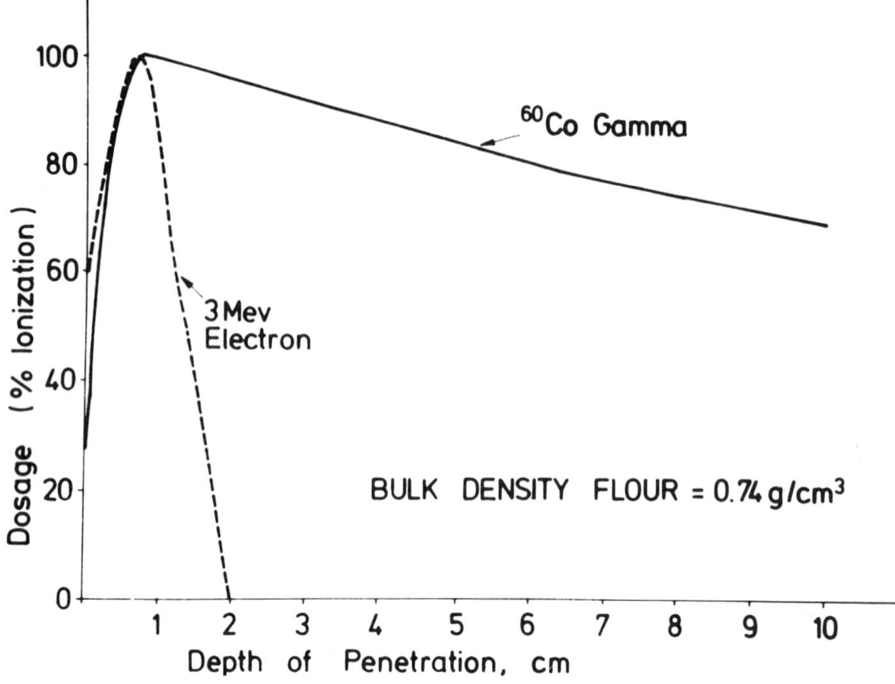

FIGURE 8. Comparison of penetration for ^{60}Co gammas and 3-MeV electrons in flour. From Welt, M. A., Radiation Sterilization — a contractors' viewpoint, presented at 76th Annu. Meet. of the Am. Soc. of Microbiol. Atlantic City, May 6, 1976. With permission.)

The useful penetration can be increased by irradiating both sides of the material. At present, health authorities in many countries approve electron energies up to a level of 10 MeV and X-ray energies only up to 5 MeV for food irradiation (see Appendix).

Machine sources such as electron accelerators have the advantage of the on/off switch and the absence of a radioactive material in the irradiation facility.

Dose rate from γ-ray sources is usually below 10 Gy/sec. Dose rate from electron accelerators is usually between 10^4 and 5×10^8 Gy/sec. Therefore, instead of the hours required of γ sources, only 1 sec or less is needed for delivery of a sterilizing dose by an electron accelerator.

III. IRRADIATION FACILITIES

The process of food irradiation shall be carried out in facilities licensed, registered, and controlled (periodically inspected) by the appropriate national authorities.[4] Licensing agencies and requirements vary from country to country.

Irradiation facilities provide for treatment of foods under controlled conditions. In any given application, the amount of radiation is controlled by knowing the rate of energy output of the source, by controlling the physical relationship (mainly distance) between the source and target material, and by controlling the time of treatment. The amount of energy absorbed is termed the "dose", which usually has been measured in rads and is now measured in grays (Gy).

The design of food irradiation facilities has been derived from actual experience with nonfood irradiations, i.e., the irradiation of medical supplies, electrical insulating and packaging materials, plastics, waste water, and sludges.

The field of radiation sources and facility design is a dynamically developing area. Facility designs are attempted to optimize the dose uniformity ratio, to ensure appropriate dose rates, and, where necessary, to permit temperature control during irradiation (e.g., for the treatment of frozen food) and also control of the atmosphere.

Depending on the design of the irradiation facility, the products can be treated in packages/ containers or they can be treated in bulk. Since radiation processing does not induce radioactivity or heat, products can be handled or shipped as soon as they leave the radiation source.

Besides the radiation source, basic components of irradiation facilities are the irradiation cell (biological shield) designed to protect personnel operating the facility from exposure to radiation, the maze and some form of carrier or conveyor system to bring the food product in proximity to the source for processing, and a safety interlock/control console system. For a given product presented in an established loading pattern and a given source size or machine output, the speed of the conveyor movement will determine the dose absorbed. An interlock system will ensure that conveyor movement occurs when the source is in the "exposure" position or the machine is switched on. Conversely, no conveyor movement shall occur when the source is in the "safe" position or a machine is "off", thus ensuring that product cannot pass through the cell without receiving treatment.[28]

A. Gamma Irradiators

The central component of a γ-irradiation facility is the radiation source. The amount of radioisotope needed depends on the amount of product to be treated per unit of time, the dose required to achieve the desired effect, and the net utilization efficiency of the radiation source.[1]

The largest ^{60}Co installations can accommodate up to 6 MCi radioactive cobalt. In large γ-radiation facilities a water filled pool usually serves as source storage and for unloading/ loading and manipulating the source elements under water during replenishment of the source. At 3.5-m depth of water 100 kCi ^{60}Co will give a dose rate approximately 0.1 mrad/hr at the surface of the pool. Products are radiation treated when the γ source is lifted up from the pool in the radiation cell. A few irradiators were designed for dry storage of radioisotopes.

The source arrangement is designed for maximum efficiency, i.e., to maximize the ratio of energy absorbed in the product to energy emitted from the source. The efficiency is dependent on the γ energy, physical source dimensions and density, irradiation geometry, product dimensions, and density.[5] Losses in efficiency are due to self-absorption in the source, absorption in nonproduct material, i.e., conveyor parts, unabsorbed radiation, and excess energy absorbed in the product above the minimum dose required for processing. There is a practical limit of efficiency of energy utilization. Measured efficiencies of commercial facilities typically range from 20 to 30%.

The maze (labyrinth) allows the products to pass the shielding without allowing radiation to escape the irradiation cell. Materials of shielding that have evolved over the years are high-density concrete, soft steel, lead, and earth. The labyrinth makes use of the property of ionizing radiation to travel in straight lines only.

Auxiliary systems are needed to deionize the water in the pool and remove excess heat, and an air handling system is needed to vent the ozone produced.

Concerning the product transport of γ irradiators, there are many kinds of constructions. They fall, however, into four categories, i.e., conveyor belt, carrier, pallet box, and bulk systems. Here only a few examples can be mentioned.

The layout of a multipurpose irradiation facility which is in operation at the IRE-MEDIRIS, Fleurus, Belgium, is shown in Figure 9. This facility, which was upgraded from a research irradiator, uses carriers (a load of 250 kg per carrier) and overlapping source plaque of ^{60}Co. It is a batch type irradiator in order to enable high flexibility in the irradiation program (different product densities, high diversity in dose requirement, etc.). It is provided with an

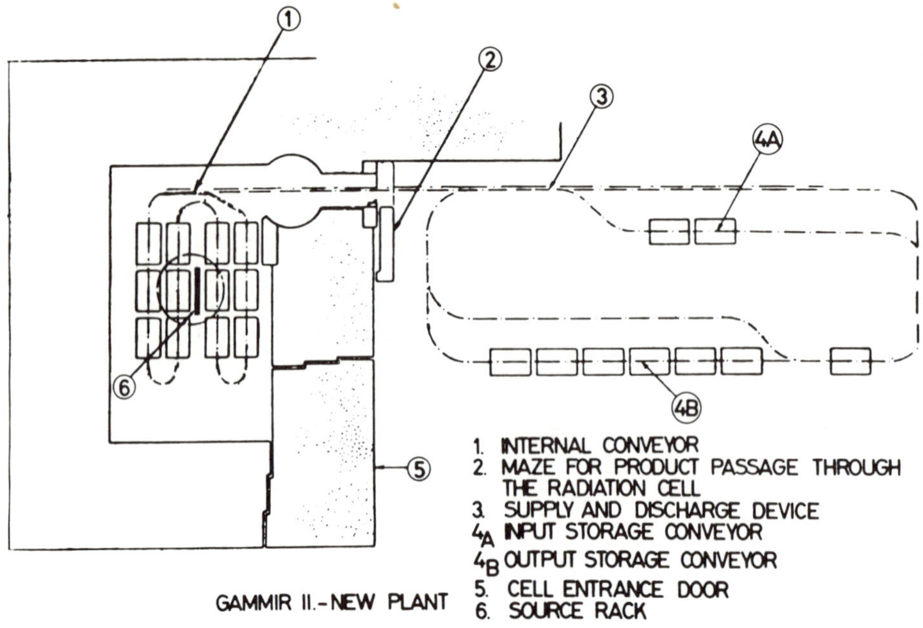

FIGURE 9. The GAMMIR II plant at the Institut National des Radioéléments, Fleurus, Belgium. (From Etienne, J. C. and Buyle, R., *J. Food Eng.* 3, 227, 1984. With permission)

automatic loading and unloading of the irradiation room, and it can unload and load a series of carriers during the irradiation of the second series of carriers.

It utilizes a 2-m-high source rack and aluminum carriers containing a 1.6-m-high product stack. Each carrier has a volume of 0.5 m^3. The carriers circulate on a closed loop. The irradiation facility was constructed to process 12 product carriers which are conveyed to the irradiation chamber on a batch basis and parked in a double row on each side of the source rack. The carrier movement around the radiation source is done automatically. The internal time setting, which depends on the installed ^{60}Co activity, density of the product, and the radiation dose required is regulated by the master timer on the control console.

All carriers are loaded on an external loading station before being evacuated to the "input storage conveyor". The "input storage conveyor" accommodates 12 product carriers awaiting automatic entry into the irradiation chamber. Following irradiation, the carriers containing the irradiated product are returned automatically to the "output storage conveyor" before being unloaded into the unloading station. After unloading, the empty carriers are automatically transferred to the loading station for reloading with unirradiated product. Loading and unloading are performed manually.

Operating for 7000 hr/year and allowing 20 min to load and unload each batch, the irradiator can process about 23,500 ton/kGy/year with a 300-kCi ^{60}Co source.

For an example of large volume, low labor cost handling, a versatile γ-irradiation facility is the JS-9000 pallet irradiator which is in operation in the Gammaster, b.v., Ede, The Netherlands (Figure 10).[7-9] The pallet concept makes it possible to irradiate product stacked on transportation pallets. This concept thus reduces the labor-intensive handling of product.

This irradiation facility is designed for radiation processing of products on pallets. Two pallets of products are vertically loaded above each other into carriers suspended from overhead rails. Four carriers are moved simultaneously around the overlapping cylindrical source so that each face of the pallets is exposed to the source twice before completing its cycle (i.e., a total of eight positions per pallet). This, coupled with strategically located attenuators, results in acceptable dose heterogeneity in spite of the large size of the unit

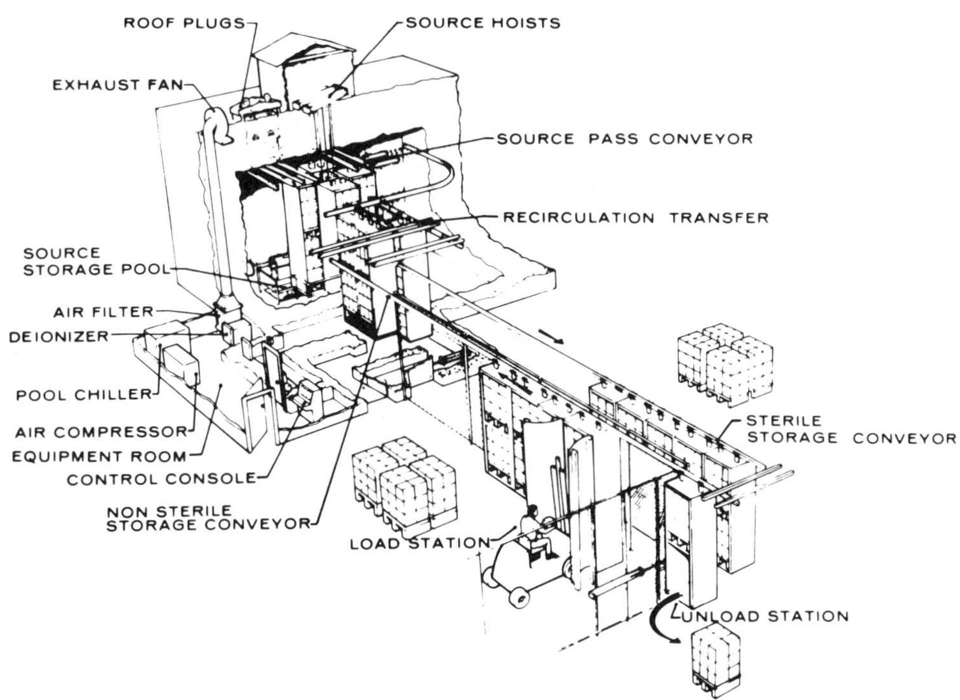

FIGURE 10. The AECL JS 9000 Pallet Irradiator commissioned by Gammaster Company, Ede, the Netherlands. (From Leemhorst, J. G., *Food Irradiation Now,* Martinus Nijhoff, The Hague, 1982, 60. With permission.)

(D_{max}/D_{min} is 1.58 to 2.14 for product densities of 0.4 to 0.8 g/cm^3). A recirculating loop makes the irradiator suitable for incremental dose application, and a specific control mechanism of each carrier permits simultaneous irradiation of products with different dose requirements.

In some applications the possibility of bulk irradiation without using bags, boxes, or trays allows a reduction of plant dimension, complexity, and material handling costs. The advantage of bulk irradiation is that a higher radiation utilization efficiency can be achieved than with package irradiators.

Several types of irradiators have been proposed for powdered or granular bulk solid material. One type of bulk product irradiator which is designed by Atomic Energy of Canada Limited and is currently used commercially to sterilize enzyme powders is shown in Figure 11.[9] The product (kept in the charge hopper) is moved by a screw conveyor into a metering chute. From this, predetermined quantities are dispensed into metal carriers within the irradiation room. The carriers move around the source until the required dose has been reached. The product is then ''dumped'' into a discharge chute and piped out of the irradiation room to the processed product storage where it is packaged.

In an Italian design, dose uniformity is assured by a pneumatical transport system into appropriate channels with a particular air mixing method known as spouted bed mixing.[10] The spouting technique has been successful for solid blending on an industrial scale. The cross section of this type of irradiation cell is shown in Figure 12. In principle, this type of facility permits simultaneous treatment of different materials, utilizing channels with different residence times.

For finely granulated materials fluidized bed systems can be designed, like the one constructed for radiation induced dry chlorination of PVC.[11]

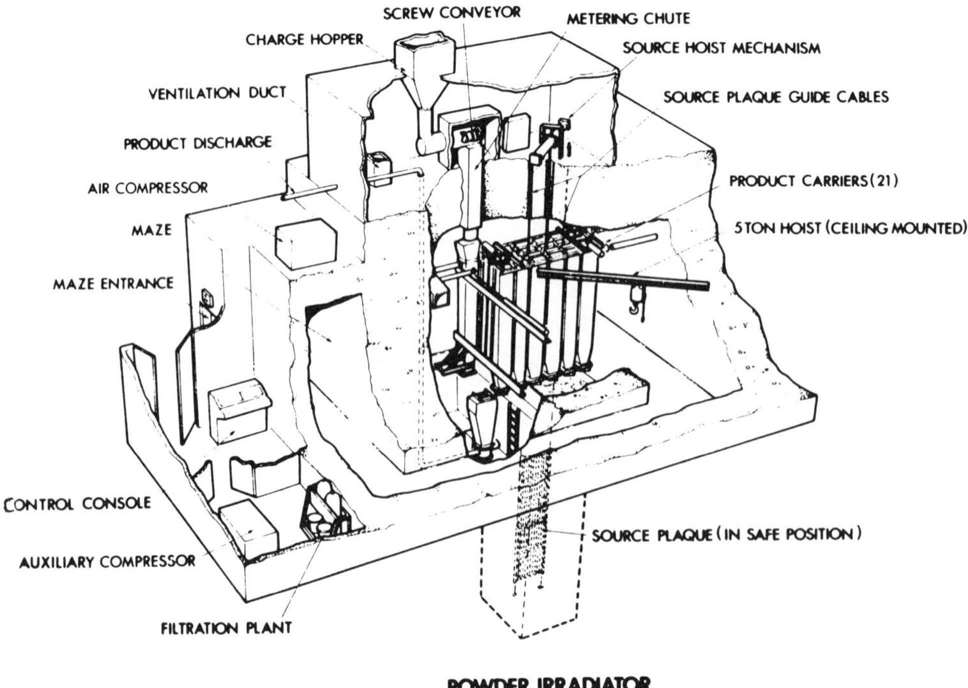

FIGURE 11. Bulk product irradiator. (From *AECL Industrial Irradiators,* Atomic Energy of Canada Limited, Radiochemical Company, Kanata, Ontario, June 1983, 13. With permission).

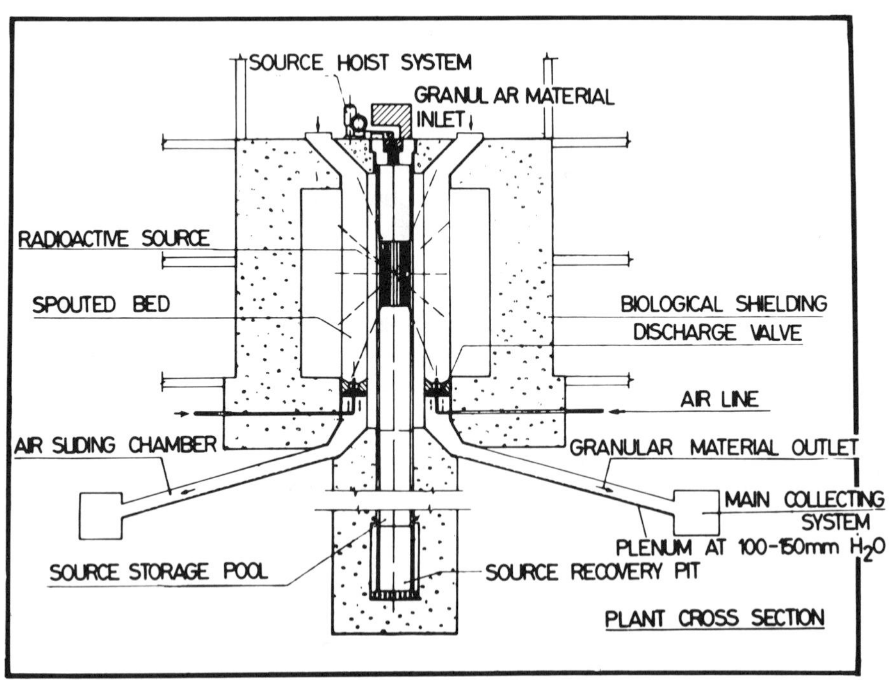

FIGURE 12. Granular materials irradiation plant with spouted bed mixing. (From Carassiti, F. and Tata, A., Proc. of the Int. Symp. on Appl. and Technol. of Ionizing Radiat., King Saud University Libraries, Riyadh, 1983. With permission).

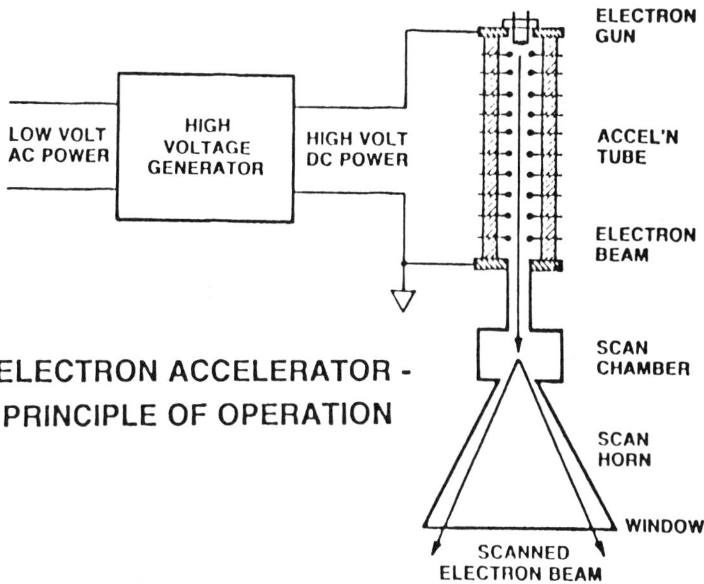

FIGURE 13. The basic concept of a direct action electron accelerator. (From Cleland, M. R., Farrell, J. F., and Morganstern, K. H., The use of high-energy electrons and X-rays for the preservation of meat and other food products, presented at the Workshop on Food Irradiat. in conjuction with the Int. Meat Res. Congr., Colorado Springs, August 31 to September 5, 1980, Tech. Inf. Ser., TIS 80-9, Radiation Dynamics, Inc., Melville, N.Y. With permission.)

B. Machine Irradiators

Electron accelerators are currently used for dozens of industrial applications including treating the insulation on wire and cable, cross-linking plastic food wrap, and curing surface coatings. The only commercial food use of machine radiation at this time is the two electron accelerator used to disinfest grain at the Odessa Port Elevator.[25,27] An industrial demonstration facility with a 75-kW electron accelerator was constructed in Israel for radiation decontamination of poultry feed.[26] Experimental work on machine sources of radiation to treat food and feed is being conducted in several countries.

The basic concept of a direct action electron accelerator is shown in Figure 13.[12] It consists of a high voltage generator connected to an electron gun, an evacuated acceleration tube, and a beam scanner. The electron stream emitted from the electron gun is injected into the accelerating tube, emerging as an intense beam. This is passed into a scan system which scans the beam back and forth before it emerges through a thin metal foil ''window''. This produces a curtain of accelerated electrons through which the product being treated is passed.

Biological shielding and venting are also required for machine sources of radiation.

Electron accelerators have specific controls for the voltage and beam operation. Direct action electron accelerators are available with voltage ratings from 0.5 to 5 million and beam power ratings from 20 to 200 kW.[13]

Indirect (microwave) accelerator techniques (linear accelerators) are also used to produce high electron energies, up to the equivalent of 10 or more million V in the beam power range of 25 to 75 kW.

The beam power ratings determine the maximum production rate. Theoretically, a 1-kW beam can process up to 360 kg/hr with an average radiation dose of 10 kGy (see Chapter 4). Practical throughput rates might be only 25 to 50% of this value depending on the product shape and configuration of the conveyor.[12]

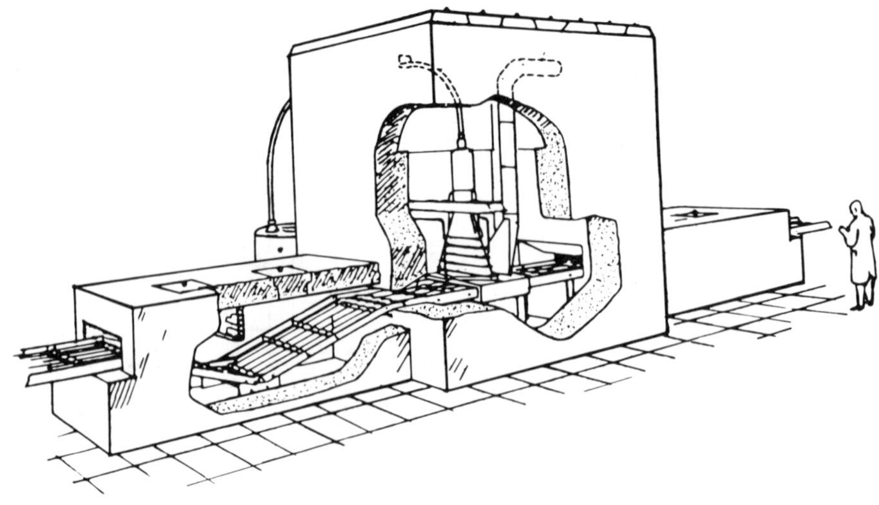

FIGURE 14. An industrial set-up of an electron accelerator. (From Oosterheert, W. F., Facilities for Radiation Processing: Principles of Isotopic Sources and Electron Accelerators, L-4, Lecture material for the IFFIT Training Course on Food Irradiation, International Facility for Food Irradiation Technology, Wageningen, 1980. With permission.)

For a typical food processing application, assuming a dose requirement of 3 kGy, double-sided irradiation, and an efficiency of 50% a throughput of 300 kg/hr can be achieved for each kilowatt of beam power. A 25-kW facility will handle 7.5 ton/hr, a 150-kW facility, 45 ton/hr.[14]

The product conveyor will be determined by the type of product to be handled. Packaged products or individual items can be handled on monorail carriers or horizontal conveyors or cart systems. Granules, powders, etc. can be handled on belt or vibratory conveyors or through pneumatic transport systems. In each case, the conveyor has to be able to operate in the special environment of the radiation cell.

Ventilation of the cell is necessary to deal with the large volume of ozone generated by the passage of the electron beam through the air.

Two typical electron irradiation set-ups are illustrated in Figures 14 and 15.[14,15]

Small, self-shielded electron accelerators at energy levels of 150 to 300 keV can be inserted directly into an existing processing line. However, these machines are capable of quite limited penetration. Manufacturers of these accelerators are studying their suitability for treating food particulates and powders.[29]

When penetrating X-rays (Bremsstrahlung) are produced, this form of radiation can be used to treat bulky objects and packages. The conversion of electron beam power to X-ray power is an inefficient process, but it can be competitive with gamma irradiation for high-capacity plants.[12]

While common X-ray machines are inefficient converters of "soft" electron energy (<0.5 MeV) to photons, the conversion efficiency increases linearly with energy, and it is believed that some electron accelerators can produce penetrating photons cheaply enough complete with ^{60}Co gammas.

The possibility exists of using either X-rays or electrons in the same facility. A proposed 6-MV, 300-kW electron accelerator, used as an X-ray generator, would have a processing capability of a 3-MC; ^{60}Co facility.[12] (The throughput rate is estimated to be about 60,000 ton/year in 6000 hr at about 5% power utilization efficiency and a 5-kGy minimum dose).

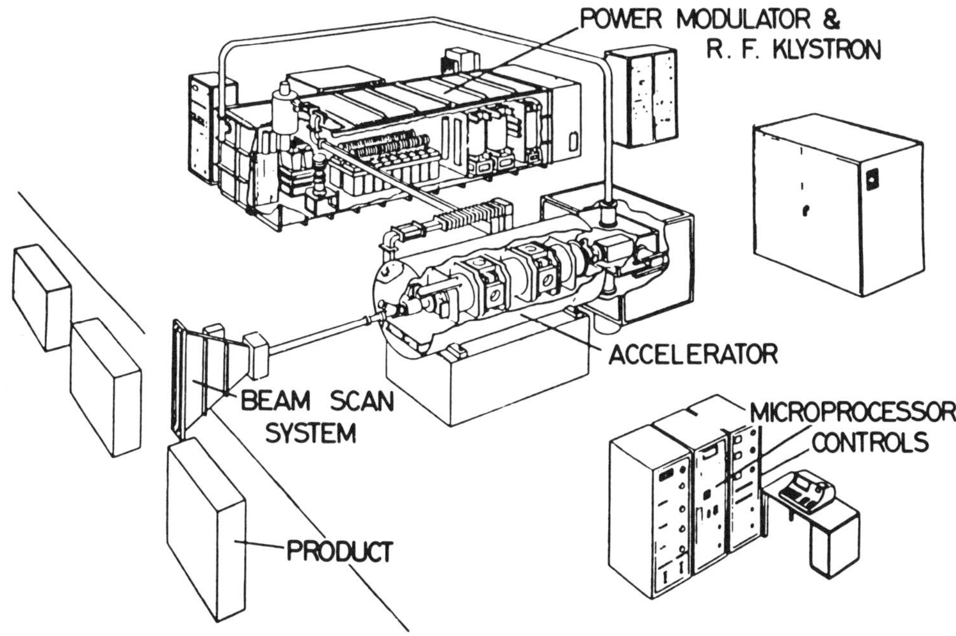

FIGURE 15. Layout of a "Dynarad" electron accelerator. (From Boaler, V. J., *J. Food Eng.*, 3, 290, 1984. With permission.)

IV. PROCESS CONTROL AND ENGINEERING ASPECTS

One of the advantages of ionizing radiation over fumigation is that radiation permits precise process control.[16,24]

Control of the food irradiation process is all types of facilities involves the use of accepted methods of measuring the absorbed radiation dose and of monitoring of the physical parameters of the process.[17] Dosimetry for food irradiation is analogous to heat penetration measurement in thermal processing with dosimeters being the analogs of thermocouples in that they are employed to provide an accurate measure of the rate of energy delivery and total energy delivered or absorbed. There are several types of dosimeters, the most common ones being based on a chemical change that is linear within a practical dose range.

Dosimetry is the keystone of proper radiation processing. Careful dosimetry is required to ensure that a technologically useful dose has been applied, while maintaining the best possible dose uniformity ratio. In addition, a knowledge of the absorbed dose is needed in order to ensure that it does not exceed the established legal dose limit.[18] Therefore, prior to the commissioning of the plant, extensive dosimetric calibrations of the irradiator are carried out and are followed during the processing by a routine dosimetry. Routine dosimetric systems are now sufficiently accurate, easy to use and reasonably priced to meet the requirements of industrial production and control. The dose distribution in the irradiated materials is monitored systematically to keep the absorbed dose between the minimum and maximum permissible values. This is easier in packaged products and γ-irradiators than in transported bulk material or in machine irradiator systems. The source geometry and mode of irradiation container or carrier movement will determine the dose distribution throughout the carrier. This can be measured directly using a "dummy" or "phantom" target, which is designed to fill a typical carrier and allow dosimeters to be distributed in a manner which will give the dose distribution at numerous points throughout the carrier, ensuring measurements at the points of minimum and maximum dose. Validation dosimetry is repeated following any major change in plant design or source size if the dose distribution in the carrier is changed.

The irradiation of unpacked pourable bulk material requires the application of a dosimeter which can be readily transported along with the material being irradiated.[19] In the case of electron accelerators the dose is mainly determined by the beam current. Coverage is ensured by the proper scan width. Quantitative measurement of dose can be made using various dyed films placed under product targets both "phantom" and actual.

Having established the convenient points at which to place dosimeters for routine dosimetry, dosimeters should be used at regular intervals in the case of a "continuous" plant. With a radiation plant operating on the "batch" principle, there must be at least one dosimeter within a batch carrier during the irradiation.[28]

In the frame of the FAO/WHO International Food Standards Programme, the Codex Alimentarius Commission has already established a Recommended International Code of Practice for the Operation of Radiation Facilities used for the Treatment of Foods (Appendix). It is required that "good irradiation practices" are followed and that doses are measured and recorded using a standardized dosimetry.

One of the great benefits of irradiation is that it gives a wide choice of packaging materials. All thermoplastic films, metal foils, and their laminates should be considered. Even in the sterilizing dose range, irradiation causes no significant change in the gas or liquid permeability or in other physical properties of most of the plastic packaging materials (see also Chapter 6, Section III).

V. OPERATIONAL AND ENVIRONMENTAL SAFETY

As in any other industrial facility, the safety of personnel is of prime concern in the irradiation facilities. Dose levels used to destroy microorganisms are several orders of magnitude higher than those killing man, and therefore, very strict safety precautions must be taken to protect workers in radiation plants.[24]

International and national regulations demand that the radiation sources are manipulated and stored in a way that excludes every possibility of radiation hazard. According to current recommendations of the International Commission on Radiological Protection, the maximum permissible doses for radiation workers (occupational limit) is a 5 rem/year (50 mSv/a), and for members of the public it is 0.5 rem/year (50 mSv/a), and for members of the public it is 0.5 rem/year (5 mSv/a). To comply with these regulations, irradiators have a well-separated and shielded area for storage of the radiation source and treatment of the products, which can be entered only when the source is in the safe position or the machine irradiator is switched off. Protection against radiation is based on the choice of suitable shielding material and wall thickness. Lead, concrete, and water are the major shielding materials.

Provisions must be made in the shielding design to allow the replenishment of the radionuclide source as it decays, for access to the irradiation zone for maintenance, and for normal flow of the irradiated product. The transport route is constructed in a way barring radiation from reaching the entrance or exit. In large irradiators isotopic sources are usually stored in a water pool several meters deep, which absorbs γ-rays effectively in order to allow human entry into the processing chamber for adjustment of conveyor systems, etc.

The control system for irradiation is designed to follow the progress of the product through the irradiation cycle, to control conveyor operation, to control the gamma source position and shielding, and to prevent and control fires. To preclude the possibility of operator exposure to radiation, interlock systems are required. Standard practice calls for double interlock systems (mechanical and electrical) to ensure that the operator is not jeopardized due to the failure of a protective device. An electronic system controls the sequence of operations, and all control functions are displayed on a control console. There is a positive indication of the correct operational and of the correct safe position of the source which is interlocked with the product movement system.

Special care is paid to the security locks to avoid unwarranted interference, to meet health, fire, and accident prevention regulations.

A safety interlock system is integrated in the control and radiation monitoring system and causes directly the lowering of the γ source into the storage position by gravity when there is a break in normal operation.

When electron accelerators are being used, the same principles are applicable. Being in the ''off'' position means that the current has been switched off and cannot accidently be turned on as long as people are inside the radiation cell.

A radiation monitoring system guards all locations where radiation could escape the radiation shield.

From an engineering point of view all this means a rigorous application of the ''fail-safe'' principle meaning that doing something wrong — even on purpose — will always result in a safe situation.

Experience has shown that there is no need for the addition of any measurable amount of radiation to the existing (natural) background radiation, i.e., the workers and neighbors at or near such facilities need not receive any more radiation from the radiation processing facilities than they do from anything else in the same general area (there is a background radiation everywhere, including within the bodies of human beings). Commercial irradiators are generally designed to prevent radiation level (dose rate) outside the irradiation chamber from exceeding 0.25 mrem/hr (2.5 μSv/hr).[6] In this way workers in a food irradiation plant would thus potentially receive a maximum annual dose during a 40-hr work week for 50 weeks/year of 500 mrem. In reality, they receive much less. (It is worthwhile to note in this context that from all natural sources we are exposed to approximately 130 mrem of radiation per year. During flying in an airplane at 10,000-m altitude the dose rate is 0.5 to 1 mrem/hr. A yearly average radiation dose from watching color TV 3 hr/day is estimated to be of approximately 6 mrem). According to risk estimations based on determination of alkyl groups in hemoglobin in persons occupationally exposed to EtO at the respiration rate of light work, and EtO exposure dose of 1 ppm/hr results in a tissue dose that is estimated to involve a risk amounting to approximately 10 mrad equiv.[20-23]

Adequate ventilation of the irradiation cell is necessary to deal with the ozone and nitrogen oxides generated by ionizing radiations in the air. The ozone concentration must be reduced to levels below the tolerance concentration of 0.1 ppm before entering the irradiation room.[1] The exhaust system also needs to provide adequate dilution of the ozone as it exhausts into the atmosphere, to ensure that the allowable levels are not exceeded in the surrounding area. (This is helped by the fact that ozone is an unstable gas, quickly decomposed to oxygen.)

One can conclude that occupational safety of personnel working in radiation installations can be very well controlled. The reliability of the gamma irradiation system is illustrated by the fact that radiation plants in the Netherlands are operating without human attendance during nights and weekends. It has also been shown that guards, gates, fences, frightening warning signs, etc. are not necessary for the safe operation of radiation facilities.

REFERENCES

1. **Brynjolfsson, A.,** Cobalt-60 irradiator design, in *Technical Developments and Prospects of Sterilization by Ionizing Radiation, International Conference, Vienna, April 1—4, 1974,* Multiscience, Montreal, 1974, 145.
2. **Seltzer, S. M., Farrell, J. P., and Silverman, J.,** Bremsstrahlung beams from high-power electron accelerators for use in radiation processing, *IEEE Trans. Nucl. Sci.,* 30(2), 1625, 1983.

3. **Welt, M. A.,** Radiation sterilization — a contractor's viewpoint, presented at 76th Annu. Meet. of the Am. Soc. of Microbiol., Atlantic City, May 6, 1976.
4. **Ballantine, D. S.,** Radiation processing and the regulatory process, *Radiat. Phys. Chem.,* 14, 245, 1979.
5. **Williams, J. L. and Dunn, T. S.,** Radiation sources — gamma, *Radiat. Phys. Chem.,* 14, 185, 1979.
6. **Etienne, J. C. and Buyle, R.,** Electro-mechanical engineering aspects in irradiator design, presented at the Semin. on Eng. Aspects of Food Irradiat., Food Engineering Forum, London, January 18, 1984.
7. **Fraser, F. M.,** Gamma radiation processing equipment and associated energy requirements in food irradiation, in *Combination Processes in Food Irradiation,* International Atomic Energy Agency, Vienna, 1981, 413.
8. **Stegeman, H.,** Progress in food irradiation — the Netherlands, *Food Irrad. Inf.* 12, 79, 1982.
9. AECL Industrial Irradiators, Atomic Energy of Canada Limited, Radiochemical Company, Kanata, Ontario, June 1983, 11.
10. **Carassiti, F. and Tata, A.,** Granular materials irradiation plant with spouted bed mixing, in *Proc. of the Int. Symp. on Appl. and Technol. of Ionizing Radiat.,* Vol. 3, King Saud University Libraries, Riyadh, 1982, 1515.
11. **Baer, M., Friese, K., and Reinhardt, J.,** A pilot plant for the radiation induced chlorination of PVC, IAEA-CN-40/29 P, presented at the Int. Conf. on Ind. Appl. of Radioisotopes and Radiat. Technol., Grenoble, September 28 to October 2, 1981.
12. **Cleland, M. R., Farrell, J. P., and Morganstern, K. H.,** The use of high-energy electrons and x-rays for the preservation of meat and other food products, presented at the Workshop on Food Irradiat. in conjunction with the Int. Meat Res. Congr., Colorado Springs, August 31 to September 5, 1980, Tech. Inf. Ser., TIS 80-9, Radiation Dynamics, Inc., Melville, N. Y.
13. **Cleland, M. R., Morganstern, K. H., and Thompson, C. C.,** High power DC electron accelerators for industrial applications, in *Proc. of the 3rd All-Union Conf. on Appl. Accelerators,* Leningrad, June 26 to 28, 1979, Tom. I.51.
14. **Boaler, V. J.,** Electron accelerator facilities for food processing, presented at the Semin. on Eng. Aspects of Food Irradiat., Food Eng. Forum, London, January 18, 1984.
15. **Oosterheert, W. F.,** Facilities for Radiation Processing: Principles of Isotopic Sources and Electron Accelerators, L-4, Lecture material for the IFFIT Training Course on Food Irradiation, International Facility for Food Irradiation Technology, Wageningen, 1980.
16. **Frohnsdorff, R. S. M.,** Sterilization of medical products in Europe, *Rev. IRE,* 5(2), 7, 1981.
17. Manual of Food Irradiation Dosimetry, Tech. Rep. Ser. No. 178, International Atomic Energy Agency, Vienna, 1977.
18. **Ladomery, L. G.,** Regulatory control of an international trade in irradiated foods, L-144, presented at the IFFIT Special Course on Public Health Aspects and Proper Control of Food Irradiation, International Facility for Food Irradiation Technology, Wageningen, 1982.
19. **Grünewald, Th. and Rudolf, M.,** Verwendung von Halbleiterdioden zur Dosimetrie bei der Bestrahlung von Schüttgut, *Atomkernenergie/Kerntechnik,* 43, 284, 1983.
20. **Calleman, C. J., Ehrenberg, L., Jansson, B., Osterman-Golkar, S., Segerbück, D., Swenson, K., and Wachmeister, C. A.,** Monitoring and risk assessment by means of alkyl groups in hemoglobin in persons occupationally exposed to ethylene oxide, *J. Environ. Pathol. Toxicol.,* 2, 427, 1978.
21. **Ehrenberg, L.,** Methods of comparing risks of radiation and chemicals. The rad equivalence of stochastic effect of chemicals, in *Radiobiological Equivalents of Chemical Pollutants,* International Atomic Energy Agency, Vienna, 1980, 11.
22. **Ehrenberg, L. and Hussain, S.,** Genetic toxicity of some important epoxides, *Mutat. Res.,* 86, 1, 1981.
23. **Latarjet, R., Averbeck, D., Levy, S., and Poirier, V.,** Quantitative comparisons of genotoxic effects of atomic energy and fossil-fueled energy. Rad-equivalences for ethylene, ethylene oxide, and formaldehyde-consequences for decisions at government level, in *Health Impacts of Different Sources of Energy,* International Atomic Energy Agency, Vienna, 1982, 387.
24. **Farkas, J.,** Radiation decontamination of dry food ingredients and processing aids, *J. Food Eng.,* 3, 245, 1984.
25. *High-Efficiency Industrial Plant for Radiation Disinfestation of Grain,* Techsnabexport, Vneshtorgizdat, Moscow, 1984.
26. **Klinger, Y., Lapidot, M., and Ross, I.,** Feed radicidation in Israel — an update, in *Food Irradiation Processing,* International Atomic Energy Agency, Vienna, 1985, 117.
27. **Zakladnoi, G. A., Men'shenin, A. I., Pertsovskii, E. S., Salinov, R. A., Cherepkov, V. G., and Krsheminski, V. S.,** Industrial application of radiation disinfestation of grain (in Russian), *At. Energ.,* 52(1), 57, 1982.
28. **Ley, F. J.,** Process control to certify the irradiation treatment of food, in *Food Irradiation: Some Regulatory and Technical Aspects,* IAEA-TECDOC-349, International Atomic Energy Agency, Vienna, 1985, 44.
29. **Aaronson, J. N. and Nablo, S. V.,** Performance characteristics and typical industrial applications of Selfshield® electron accelerators (300 kV), *Nucl. Instr. Methods Phys. Res.,* B10/11, 998, 1985.

Chapter 4

ECONOMIC FEASIBILITY OF RADIATION PROCESSING OF DRY INGREDIENTS

Like other physical methods of processing food, irradiation technique involves relatively
... minimum capacity of the irradiation facility for
... the simpler type of ethylene oxide (EtO) facility
...on plant. The capital cost of radiation processing
... the type of irradiation desired and the capacity
... facility would cost \$500,000 (pilot plant) to \$4
... and the annual operating expenses including am-
...on. However, unlike alternative processes, irradia-
...erefore, in a given plant, the cost of irradiation is
...se requirement.

... both radiation and EtO processes on a commercial
...lucts under generally comparable conditions in the
...ess was about double that for irradiation, without
... challenge testing that is required.[1]

... of an irradiator, there is a marked drop in processing
...g capacity, and the utilization of the irradiator has
...osts.

...n source size, power in kilowatts, dose requirement

$$\ldots 0 \frac{f \cdot W}{D} \text{ kg/hr} = 7920 \frac{f \cdot W}{D} \text{ lb/hr}$$

...n which the emitted radiation power is absorbed, W
...e, and D is the dose in kilograys required to achieve
...r 312,000 Ci of ^{137}Cs emit 1 kW of γ-radiation).
...ate the source size requirement in terms of kilowatts

$$\ldots = \frac{1}{3600} \frac{X \cdot D}{f} \text{ kW}$$

...e irradiated per hour at D kilograys.

...ortional to the dose requirement, for economic reasons
...nt reduction in the viable cell count instead of complete
...es described earlier (see Chapter 2, Section VII) the
... the products as applied in the food processing industry
...on decontamination of their dry ingredients.

...edients with a dose requirement of 4 to 8 kGy is estimated
...xisting Dutch ^{60}Co γ-irradiation service plants.[3,4] This is approximately 1 to 8% per... of the current price of spices. Similar cost is given for ^{60}Co irradiation of fish protein concentrate by an author from the U.S.[5] Canadian estimates gave \$0.10 to \$0.15/kg as the processing cost for reduction of the microbial load of spices considering a dose requirement up to 10 kGy.[6,7] Hungarian authors estimated that radiation decontamination of ground paprika and its treatment with EtO would cost about the same.[8]

It should, however, be emphasized that processing costs depend very much on throughput and total plant utilization, which once again will depend on special circumstances. γ-

Irradiators emit radiation 24 hr/day; therefore, it is desirable to operate as many hours per day or per week or per year as possible for the most favorable economics.

Assuming the applicability of high-capacity machine irradiation, Cleland et al.[9] estimated a 0.7-cent/kg processing cost for direct electron treatment at a minimal dose of 5 kGy and a yearly throughput of 100,000 tons. Using the electron accelerator as an X-ray generator, the same authors estimated that the unit processing cost would be 1.5 cent/kg (also at 5 kGy) assuming a yearly throughput of 60,000 ton. Unit cost would be somewhat higher for a smaller plant since the capital investment does not scale down in proportion to the source capacity.

Recently Deitch[10] published useful detailed tables which may be used for estimating the costs of irradiation on the basis of numerous components of the investment and the operational costs, both for a 10-MeV linear accelerator and a ^{60}Co facility. Deitch was using the functions for cost estimation of the individual components as developed by Brynjolfsson.[2]

One has to keep in mind, however, that the actual cost of food irradiation, as the costs of any other food processing technique, may be greatly affected by local circumstances. Due to varying conditions, costs of the same type of treatment may differ from place to place, and a cost estimate made in one country is not necessarily representative of other countries.

In spite of the fact that ionizing radiation is at present an expensive form of energy, in view of several factors outlined below it is believed that radiation treatment could compete favorably in many cases from the point of view of economic feasibility with alternative decontamination processes.[17]

1. Unlike fumigation, radiation treatment is easy to automate, and it can be applied as a continuous process.
2. Radiation can be applied to prepackaged materials in their final packaging.
3. Irradiation of ingredients need not be seasonal, but could be done practically the whole year.
4. The dose requirement to achieve practical decontamination is a moderate one.
5. The price of the dry ingredients, herbs, and enzyme preparations is relatively high, so the cost of radiation contributes only a small fraction of the cost of the final product.
6. Because of compactness, high value, and good transportability of many dry ingredients, their radiation treatment can be executed in centralized facilities and/or on a service basis, e.g., in the already existing irradiation (pilot) plants.
7. Demand for decontaminated ingredients is increasing, and, therefore, achieving a good microbiological quality may justify a higher price.

The benefits of using irradiated ingredients, e.g., in the canning and meat processing industry, have been demonstrated by several experiments when irradiated ingredients resulted in a reduced heat-treatment requirement and/or better microbiological quality of the processed foods[11-13] (see also Chapter 2, Section IX).

In addition to its cost effectiveness, radiation processing is less energy consumptive than other decontamination techniques, an increasingly encouraging feature of radiation decontamination.[14,15]

It is virtually certain that regulatory agencies, even if they do not ban EtO treatment, will restrict significantly the tolerance levels on residues of EtO and its reaction products and drastically reduce the established limit of exposure to EtO for workers in fumigation plants.[16] All of these will lead to changes in the operating practices of EtO plants, to increasing operating costs, and should further emphasize the advantages of the radiation alternative.

REFERENCES

1. **Frohnsdorff, R. S. M.,** Sterilization of medical product in Europe, *Rev. IRE,* 5(2), 7, 1981.
2. **Brynjolfsson, A.,** Factors influencing economic evaluation of irradiation processing, in *Factors Influencing the Economical Application of Food Irradiation,* International Atomic Energy Agency, Vienna, 1973, 13.
3. **Leemhorst, J. G.,** Industrial application of food irradiation, in *Food Irradiation Now,* Martinus Nijhoff, The Hague, 1982, 60.
4. **Oosterheert, W. F.,** private communication.
5. **Tsuji, K,** Low-dose cobalt-60 irradiation for reduction of microbial contamination of raw materials for animal health products, *Food Technol.,* p.48, February 1983.
6. **Ouwerkerk, Th.,** An Overview of the Most Promising Industrial Applications of Gamma Processing, Atomic Energy of Canada Limited, Industrial Products, Ottawa, March 1982.
7. **MacQueen, K. F.,** Report on Food Irradiation, Agriculture Sector, Natural Resources Division, Canadian International Development Agency, Toronto, Contract 84-72/C144, June 1, 1984.
8. **Szabad, J. and Kiss, I.,** Comparative studies on the sanitising effects of ethylene oxide and of gamma radiation in ground paprika, *Acta Aliment.,* 8, 383, 1979.
9. **Cleland, M. R., Farrell, J. P., and Morganstern, K. H.,** The use of high-energy electrons and x-rays for the preservation of meat and other food products, presented at the Workshop on Food Irradiat. in conjunction with the Int. Meat Res. Congr., Colorado Springs, August 31 to September 5, 1984, Tech. Inf. Ser., TIS 80-9, Radiation Dynamics, Inc., Melville, N.Y.
10. **Deitch, J.,** Economics of food irradiation, *Crit. Rev. Food Sci. Nutr.,* 17, 307, 1982.
11. **Dutova, W. N., Kardashev, A. V., and Gotfarsh, M. M.,** The use of gamma irradiation in cold sterilization of salt and spice mixtures. *All-Union Res. Mar. Fish. Oceanogr., Moscow.* 73, 69, 1970; English translation, Fisheries Research Board, Canada.
12. **Farkas, J.,** Radurization and radicidation of spices, in *Aspects of the Introduction of Food Irradiation in Developing Countries,* International Atomic Energy Agency, Vienna, 1973, 43.
13. **Kiss, I. Farkas, J., Ferenczi, S., Kálmán, B., and Beczner, J.,** Effects of irradiation on the technological and hygienic qualities of several food products, in *Improvement of Food Quality by Irradiation,* STI/PUB/ 370, International Atomic Energy Agency, Vienna, 1974, 158.
14. **Silberstein, O., Kahan, J., Penniman, J., and Henzi, W.,** Irradiation of onion powder: effects on taste and aroma characteristics, *J. Food Sci.,* 44, 971, 1979.
15. **Trägardh, C. and Hallström,B.,** Energy Analysis of Selected Food Post-Harvest and Preservation Systems, Irradiation Compared to Conventional Food Preservation Methods, Res. Contract No. 2850/TC, report to the International Atomic Energy Agency, Vienna, 1981.
16. Occupational Safety and Health Administration, Occupational exposure to ethylene oxide, *Fed. Regist.,* 49(122), 25734, 1984.
17. **Farkas, J.,** Radiation decontamination of dry food ingredients and processing aids, *J. Food Eng.,* 3, 245, 1984.

Chapter 5

WHOLESOMENESS OF IRRADIATED DRY INGREDIENTS

I. INTRODUCTION

The term "wholesomeness" which is applied to irradiated foods has acquired the meaning of nutritional quality, toxicological safety, and microbiological safety. Obviously, adequate proof of safety of irradiated food must exist before the health authorities can permit it to be sold to the public. No other method of food preservation has been so thoroughly scrutinized as food irradiation, and the safety of consuming foods irradiated under proper processing and handling has been well established in the past decades by very elaborate wholesomeness-testing programs.[1-3] Such studies were begun in the U.S. as early as the late 1940s.

Irradiation with energy levels foreseen for food processing (up to 5 MeV for γ- and X-rays and up to 10 MeV for electrons) does not produce radioactivity in the foods so treated.[4]

II. CHEMICAL CONSIDERATIONS

As the application to food of ionizing radiation from radioisotopes or from machine sources represents an input of energy into the food, efforts have been directed at identifying and quantifying radiolytic products formed in food thus treated. An understanding of the nature and quantity of products formed in food following irradiation is thought to be necessary to establish the safety of irradiated foods in conjunction with nutritional and biological testing.[4]

Ionizing radiations produce ions and other chemically excited molecules in the exposed medium. The activated molecules are responsible for the beneficial antimicrobial action of irradiation. However, the excited molecules will lead to some chemical and biochemical changes. The extent of these changes, among other factors, is dependent on the radiation dosage. The energy taken up by irradiated food is far less than that taken up by heated food. Thus, radiation treatment produces only very small amounts of reaction products. In various spices irradiated with a dose of 45 kGy, gas chromatographic studies revealed radiation-induced compounds at a level less than 0.01% of the total volatile constituents.[5]

Water forms a primary set of radicals, i.e., OH radicals, H atoms, and hydrated electrons. Such radicals are responsible for the indirect action of radiation in foods, because these primary radicals react with food components to form a second set of radicals and ultimately final radiolysis products.

The major factors in influencing product yields are the composition, the physical state, and the processing parameters of foods. These factors are not entirely independent of each other. The physical state of the food being irradiated (frozen vs. fluid or hydrated vs. desiccated) influences both the primary and secondary reactions and consequently affects the yields of final products.[6]

Research on product formation and reaction mechanisms have shown that despite the complexity of the primary radiolytic events, only a few key radicals are ultimately formed and become precursors of major final products.[6]

Since the ultimate fate of radicals formed depends on their ability to migrate or to twist into position for bimolecular reaction, factors that alter their environment can influence their pathways to decay and hence their lifetimes. For nonlipid components, the presence of water is of particular importance.

The state of hydration is important both in radiolysis and product quality. The presence of nonfrozen water directly increases the formation of radiolytic products. In dehydrated systems the reactions are limited due to the lack of water for primary radical formation. The influence of water on migration of other reactive entities (e.g., oxygen) is also important. Thus, in general, dry or semidry foods can take higher doses without deleterious effect to product quality.

It was demonstrated with dry grains, powdered foods, and various dry ingredients that as compared to those in moist systems, the decay of free radicals in dry products is much slower, though considerable differences could be observed among these products in changes of residual radicals during storage.[7-9,53] Therefore, in dry substances the presence of free radicals can be reckoned with as a result of irradiation (see also Chapter 6, Section IV).

The formation as well as the rate of decay of free radicals is highly influenced by the moisture content of dry ingredients as has been demonstrated with irradiated starches and spices.[10,11]

It should be pointed out that free radical formation is not limited to irradiated foods. Although the generation and subsequent reaction of free radicals comprise the major route by which radiolytic products are formed, such reactions are also common during conventional food processing and storage operations. Electron spin resonance measurements have shown that the milling of grain and the heating of proteins having a low water content also produce long-lived free radicals.[12,13] Irradiated dry ingredients are generally used in other foods that contain water. Thus, the manner in which these commodities are used provides mostly a means for rapid dissipation of the free radicals, thereby precluding their ingestion.

Notwithstanding the fact that free radicals trapped in dry substances vanish again on soaking, a long-term feeding study was especially designed to look for possible effects of an irradiated diet containing a high free radical concentration. The results have been negative both with regard to chronic toxic effects and tumor formation and to mutagenic effects in a dominant lethal assay.[14,15]

In Hungarian studies, no notable changes were found in the lipid composition of several spices and spice mixture (see Section III) after radiation treatment up to 15 kGy.[54] No significant effect could be found by thermal analysis either.[55]

Radiation chemistry studies most directly relevant to wholesomeness of dry vegetable ingredients were those performed in France with starches. It could be demonstrated that none of the radiolytic products was present in an amount to be toxic and that all of them could also be produced by other physical processes of food preservation.[3,16] Starches of different origin, i.e., maize, wheat, rice, potato, etc., contained radiation-induced radicals and radiolytic products of the same nature and order of magnitude, independent of the origin of the investigated starch.

Irradiation of starch causes partial depolymerization and fragmentation into simpler molecules (e.g., glucose and maltose as well as maltotriose and maltopentose). Formic acid, acetaldehyde, and formaldehyde are the major radiolytic products.[17,18] Product distribution, however, may differ with different water activities, e.g., in starch the formation of malondialdehyde and formic acid was reported to increase upon reducing the water content, while other reactions, i.e., the formation of glucose and formaldehyde, were practically independent of the water content.

Gas chromatograms from extracts of irradiated potato starch (Figure 16A) show that rather high doses are necessary to produce amounts of products which can be measured in a reproducible way. In the presence of other food constituents, competition will occur, and the yield of products from carbohydrates will be even more suppressed (Figure 16B). The produced compounds are found in many natural products and in conventionally treated foods. Although some of these compounds are toxic to unicellular organisms, they are not toxic to mammals in the concentration produced by irradiation.

III. NUTRITIONAL EVALUATION

Minor food ingredients usually do not play any significant role as nutrient supplies.

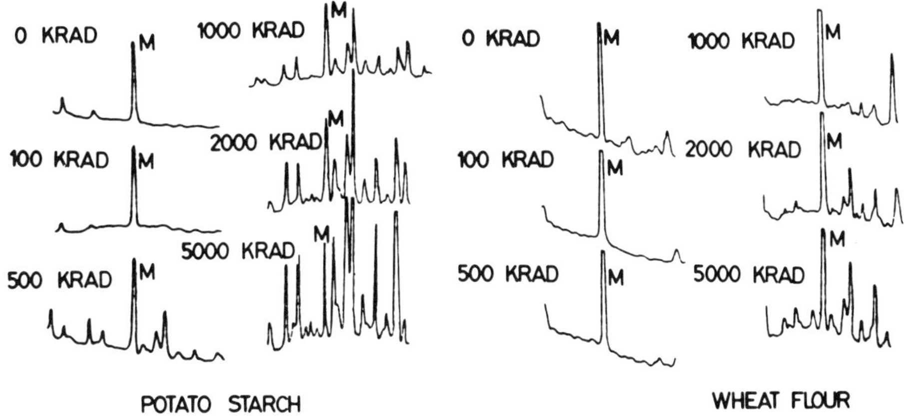

FIGURE 16. Gas chromatograms of extracts of unirradiated and irradiated potato starch and wheat flour. Extraction with ethyl acetate-acetone-water 4:5:1; reduction with KBH_4; trimethylsilylation M, mesoerythritol added as internal standard. (From Delincée, H., Radiation Chemistry of Food Components, L47 lecture handouts at IFFIT Training Courses on Food Irradiation, International Facility for Food Irradiation Technology, Wageningen. With permission.)

Nevertheless, some studies in relation to the nutritional quality of irradiated dry ingredients can be mentioned here.

Tsuji[20] reported that using a microbiological assay method, the relative nutritional value of the irradiated (10 kGy) fish protein concentrate was not reduced as compared to nonirradiated control.

The effect of irradiation on the vitamin content of whole egg powder and some cereal products should be considered, however. It is known that thiamin is the most radiosensitive of the water-soluble vitamins, and vitamin E is the most radiosensitive of the lipid-soluble vitamins.

Experiments by Diehl have shown that on storage of wheat flour and rolled oats the irradiated samples lost more thiamin, irradiated egg powder, and vitamin A than the untreated samples.[19,21,22,52] For example, egg powder irradiated with 10 kGy in the presence of air and stored for 4 months at ambient temperature lost 68% of its thiamin content. These aftereffects, which could even be enhanced by a subsequent heat treatment, are caused probably by some oxidation products present in the irradiated samples. They have not yet been fully identified.

An important finding is, however, that exclusion of oxygen in packaging, e.g., under nitrogen stream, or vacuum, greatly reduced these aftereffects in vitamin E, thiamin, and vitamin A.[52,59] Under such conditions only slight losses were found, e.g., in rolled oats (Figure 17) or in egg powder.[52,59] Exclusion of oxygen reduced also the loss of α-tocopherol, e.g., in rolled oats irradiated with 1 kGy and stored for 8 months. The loss was diminished from 56 to 5%. If the rolled oats were stored and heated, again the protective influence of the oxygen-free atmosphere was demonstrated. Subfreezing irradiation temperatures considerably improved the retention of thiamin and vitamin A in whole egg powder.[52]

Using defatted, radiation-decontaminated fish protein concentrate instead of untreated ones in compressed pet food tablets did not influence the stability of vitamins in the product.[19] No reduction in vitamin C content of radiation-decontaminated onion powder was observed.[23]

Irradiation of certain dry vegetables seems to have even some positive nutritional consequences. Radiation treatment improved the nutritive and metabolizable values of lentil as observed on the basis of chick bioassay test.[24] Presumably by reduction of cooking time, irradiated red gram *(Cajanus cajan)* shows better retention of some of the B vitamins on cooking. Slight degradation of proteins due to irradiation occurs, and hence susceptibility to proteolytic action in vitro is increased.[25]

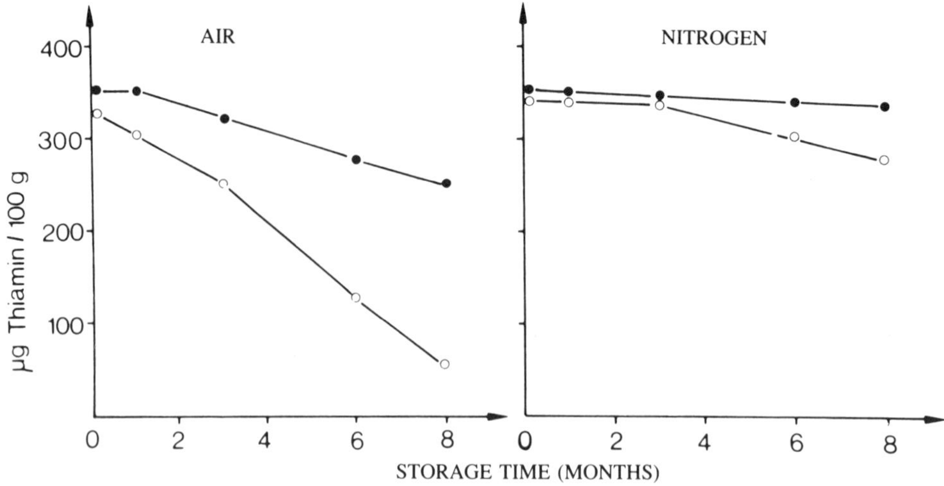

FIGURE 17. Effect of packaging atmosphere on the loss of thiamin in rolled oats upon irradiation and storage; ●----● unirradiated, ○----○ 0.25 kGy. (From Delincée, H., Radiation Chemistry of Food Components, L47, Lecture handouts at IFFIT Training Courses on Food Irradiation, International Facility for Food Irradiation Technology, Wageningen. With permission.)

IV. BIOLOGICAL STUDIES

Among a large number of representative irradiated food items, dried products, i.e., several dried fruits, flour, dessert powder, dried eggs, and powdered milk, were tested in the U.S. in systematic toxicological studies during the period of 1954 to 1964.[26] In 1964, when these studies were completed the Surgeon General's scientists concluded that "Food irradiated up to absorbed doses of 5.6 megarads with Co-60 source of gamma radiation or with electrons with energies up to 10 million volts have been found to be wholesome, i.e., safe and nutritionally adequate."[27]

Spices have been accorded a high priority in the wholesomeness testing program of the International Project in the Field of Food Irradiation (IFIP), a cooperation in which 25 countries eventually took part under the aegis of the Food and Agriculture Organization (FAO) and International Atomic Energy Agency (IAEA) between 1971 and 1981.[28] These studies involved among others irradiated wheat flour, starches, spices, and dried dates. Specific animal feeding tests as well as teratogenicity and genotoxicity studies with various irradiated dry ingredients were executed, mainly in Hungary, France, The Netherlands, Belgium, and the Federal Republic of Germany within the frame of IFIP.

In France, animal feeding studies have been performed with reassuring results not only with irradiated starches but also with mixtures of radiolytic products extracted from irradiated starches and administered to rats in concentrations many times higher than they occurred in irradiated starches.[29-31]

In the wholesomeness testing of spices, herbs, and dry vegetable seasonings we face the difficulty that feeding tests with hot and pungent materials or those containing high concentrations of biologically active compounds, i.e., those found in many of the important spices and herbs, are extremely difficult or even impracticable to perform, particularly at the high levels required in the past in the protocols for irradiated staple foods. There are exceptions to this, e.g., the powder of sweet (capsaicin-free) paprika. In Hungarian experiments this product was consumed by albino rats at a 35% level of the total diet for months without adverse effects. A subacute wholesomeness test was performed then without any difficulty with irradiated (15 kGy) and/or heat-treated (autoclaved) paprika powder at the same 35% level in the diet.[32,33] In spite of the above-mentioned difficulties, feeding studies in rats fed a diet containing a spice mixture at levels ranging from 2 to 25% have been completed.[34-36] Apart from the lower growth rate caused by the lower food consumption by rats of feed containing high levels of spice, the data show no evidence of adverse effects from feeding of spices whether irradiated (15 kGy), unirradiated, or treated with ethylene oxide (EtO) even up to a 25% dietary level. Nevertheless, it was felt that in the course of a long-term study, feeding high levels of spice would produce underweight animals and this might lead to further complications. Therefore, further research programs included short-term tests on mutagenicity and teratogenicity and the identification of some radiation-induced chemical changes in individual spices.

For teratogenicity studies, groups of pregnant CFY albino rats were fed either stock ration (controls) or diets containing either nonirradiated or irradiated (15 kGy) spices at levels of 3.5, 25, and 25% for ground black pepper, mild paprika, and spice mixture, respectively.[37] Between 10 and 15 dams per group were used, and the test diets were made avaliable to the animals from the 6th to the 15th days of gestation only. This first experiment was conducted within 2 weeks of the irradiation treatment of the spices. A similar study was conducted in which the irradiated spice had been stored for 3 months. The food consumption and growth of the dams were similar in all groups. The dams were killed on day 20 of gestation, and the contents of the uteri were examined. No treatment-related effects were reported in the numbers of implantation sites, resorptions (which were low), or viable fetuses or in the number of gross, soft-tissue, or skeletal abnormalites in the offspring. Similar findings were obtained in the experiment with the stored irradiated spices. It was concluded from these studies that no teratogenic effects were found to be associated with the consumption of diets containing either irradiated or nonirradiated black pepper, mild paprika, or spice mixture. (The spice mixture was made up of 55% paprika, 14% black pepper, 9% coriander, 9% allspice, 7% marjoram, 4% cumin and 2% nutmeg).

Cytogenetic effects of irradiated paprika were studied with micronucleus tests in mice.[58] Groups of 5 female Swiss mice were fed for 12 days on diets containing 0 (control), 20 (dry weight) paprika (capsaicin free) or 20% (dry wieght) irradiated paprika (capsaicin free). A further group of animals was fed on control diet and at 30 and 6 hr before killing was given 100 mg/kg hycanthone methane sulfonate (positive control). At the end of the feeding period the animals were killed and bonemarrow preparations made. No differences were observed between the control and paprika-fed groups in the incidence of micronucleated erythrocytes. The sensitivity of the test was confirmed by the positive findings with hycanthone methane sulfonate.

The possible mutagenic effects of irradiated paprika (capsaicin free), black pepper, and the spice mixture mentioned above were studied using the Ames test and host-mediated assays as well.[38,39] Extracts were prepared from heat-treated paprika or irradiated (50 kGy) paprika by solvent extraction with dimethyl sulfoxide (DMSO). Samples of both paprika samples and their DMSO extracts were subjected to bacterial mutagenicity tests using *Salmonella typhimurium* strains TA1535, TA98, and TA1975. No difference in the incidence of revertants was reported between the paprika samples and their extracts or between the heat-treated and the irradiated paprika. In a further study to compare the mutagenic potential of the heat-treated paprika with irradiated (50 kGy) paprika, groups of male CFLP albino mice were fed for 1 week on diets containing 0 (control), 35 heat-treated paprika, or 35% irradiated (50 kGy) paprika. A host-mediated assay was conducted using *S. typhimurium* TA1530 as the test organism. There was a high variability in the results observed though no pattern emerged to indicate any treatment-related effect. In neither study were the results indicative of a mutagenic effect on the part of irradiated paprika.

Extracts were prepared from a spice mixture, irradiated with 15 or 45 kGy, by solvent extraction with DMSO and subjected to bacterial mutagenicity tests using *S. typhimurium* strains TA1537, 1538, and 98 with metabolic activation (S9). No effects were observed with either sample. In a further study, groups of albino rats were fed for 6 days on diets containing 0 (control), 25% nonirradiated spice mixture, 25% irradiated (5 kGy) spice mixture, 25% irradiated (15 kGy) spice mixture, 25% nonirradiated paprika, 25% irradiated (5 kGy) paprika, 25% irradiated (15 kGy) paprika, 3.5% nonirradiated black pepper, 3.5% irradiated (5 kGy) black pepper, or 3.5% irradiated (15 kGy) black pepper. Samples of urine collected from these rats were subjected to bacterial mutagenicity tests using *S. typhimurium* strains TA1537, 1535, 1538, 98, and 100. No treatment-related effects were reported in the results of these tests. The results of these studies indicate a lack of mutagenic potential not only on the part of various irradiated spice extracts but also on the part of their urinary metabolites in rats.[40] Prophage induction tests for potential carcinogens using blood samples of the same animals were also performed with negative results.[41] None of these experiments has indicated induction of, or increase in, reverse mutations or other DNA-damaging effects due to irradiated spices.

In Germany, Münzner and Renner[42] investigated irradiated onion powder using the *Salmonella*/mammalian microsome mutagenicity assay and the in vivo sister chromatid exchange test. The onion powder irradiated up to 13.6 kGy did not reveal any mutagenic activity. In the U.S., onion powder irradiated with 9 kGy was found to be nonmutagenic in recessive sex-linked lethal tests in *Drosophila*.[43,44]

Investigations in Hungary on the possible genotoxic effects of irradiated onion powder by means of both Ames test and prophage induction (Inductest) also gave negative results.[45,46] Powdered garlic was irradiated at a dose of 10 kGy, and water extracts were examined in the *Salmonella* reversion test using TA98, 100, 1535, and 1537 with and without S9 mix. No mutagenic activity was noted.[47]

Results of wholesomeness studies designed to test individual food items are greatly strengthened by observations on enormous numbers of laboratory animals maintained on irradiated stock diets, where 100% of the feed intake is radiation treated, frequently at 20- to 30-kGy dose levels. Radiation processing of diets for specific stocks of laboratory animals (specific pathogen and germ-free populations) has long been a commercial practice in several countries, because irradiated feeds have clear-cut advantages over heat-treated or fumigated feeds with regard to palatability, nutrient value, and microbiological safety of the diet.[3,48-51] Already in 1979, in the U.K. alone 1200 ton of laboratory animal diets were irradiated.[48]

V. EVALUATION OF WHOLESOMENESS STUDIES

The results of complex wholesomeness studies on irradiated food, including, of course, not only the ingredients but all important staple foods, were evaluated in 1964, 1969, 1976, and 1980 by an International Expert Committee (JECFI) jointly convened by FAO, IAEA, and the World Health Organization (WHO). In 1980 a complete review of all wholesomeness studies was compiled by IFIP, Karlsruhe and was reported to the Joint Expert Committee. The JECFI concluded in November 1980 that foods irradiated at doses up to 10 kGy overall average dose present no toxicological hazard, and therefore, no further toxicological testing of such irradiated food should be necessary.[4] In addition, it was concluded that irradiated foods do not pose specific microbiological and nutritional problems.

REFERENCES

1. **Barna, J.,** Compilation of bioassay data on the wholesomeness of irradiated food items, *Acta Alimen.*, 8, 205, 1979.
2. **Diehl, J. F.,** Irradiated foods — are they safe? in *Impact of Toxicology on Food Processing,* Ayres, J. C. and Kirschman, J. C., Eds., IFT Basic Symp. Ser., AVI Publishing, Westport, Conn., 1981, 286.
3. **Saint-Lebe, L., Raffi, J., and Henon, Y.,** Le Traitement Ionisant des Denrees Alimentaires, efficacité et Absence de Risques pour l'Homme, Rapport CEA-R-5162, Commissariat a l'Energie Atomique, Service de Documentation, Gif-Sur-Yvette, France, 1982.
4. *Wholesomeness of Irradiated Food,* Report of a Joint Food and Agriculture Organization/International Atomic Energy Agency/World Health Organization Expert Committee, World Health Organization Tech. Rep. Ser. No. 659, World Health Organization, Geneva, 1981.
5. **Tjaberg, T. B., Underdal, B., and Lunde, G.,** The effect of ionizing radiation on the microbiological content and volatile constituents of spices, *J. Appl. Bacteriol.*, 35, 473, 1972.
6. **Taub, I. A.,** Reaction mechanisms, irradiation parameters, and product formation, in *Preservation of Food by Ionizing Radiation,* Vol 2, Josephson, E. S. and Peterson, M. S., Eds., CRC Press, Boca Raton, Fla., 1983, 125.
7. **Diehl, J. F. and Hofman, S.,** Electronspinresonance studies on radiation-preserved foods. I. Influence of radiation dose on spin concentration, *Lebensm. Wiss. Technol.*, 1, 19, 1968.
8. **Hayashi, T., Kawakishi, S., Namiki, M., Nara, S., and Komiya, T.,** Formation and disappearance of the radiation induced radicals in foods, *Food Irradiat., Jpn.*, 7(1), 15, 1972.
9. **Bachman, S., Galant, S., Gasyna, Z., and Witkowski, S.,** Effects of ionizing radiation on gelatin in the solid state, in *Improvement of Food Quality by Irradiation,* International Atomic Energy Agency, Vienna, 1974, 77.
10. **Beczner, J., Farkas, J., Watterich, A., Buda, B., and Kiss, I.,** Study into the identification of irradiated ground paprika, in *Proc. Int. Colloq. Ident. Irradiat. Foodstuffs,* Commission of the European Communities, Directorate - General Scientific and Technical Information and Information Management, Luxembourg, 1974, 255.
11. **Komiya, T. and Nara, S.,** The effects of moisture contents on the formation of radicals in various starches by γ-irradiation. II. On the changes of radical contents of irradiated various starches in adsorption and desorption of moisture, *Food Irradiat. Jpn.*, 9(1), 62, 1974.
12. **Redman, D. G., Axford, D. W. E., Elton G. A. H., and Brivati, J. A.,** Mechanically produced radicals in flour, *Chem. Ind. (London),* 1298, 1966.
13. **Uchiyama, S. and Uchiyama, M.,** Free radical production in protein-rich foods, *J. Food Sci.*, 44, 1217, 1979.
14. **Renner, H. W., Grünewald, Th., and Ehrenberg-Kieckebush, W.,** Mutagenicity test of irradiated foods by dominant-lethal assay, *Humagenetik,* 18, 155, 1973.
15. **Renner, H. W. and Reichelt, D.,** On the wholesomeness of high concentrations of free radicals in irradiated foods, *Zentralbl. Vet. Med.*, B20, 648, 1973.
16. **Raffi, J. and Saint-Lebe, L.,** Radiochimie des Amidons, Bilans des Travaux Realisés entre 1975 et 1980, ICS/IF/80.2a, CEN Cadarache, Saint-Paul-lez-Durance, France, 1980.
17. **Dauphin, J. F., Athanassiadis, H., Berger, G., and Saint-Lebe, L.,** Présence d'acide formique dans l'amidon de mais irradié, *Stärke,* 26, 14, 1974.
18. **Berger, G., Dauphin, J. F., Athanassiadis, H., Saint-Lebe, L., Angel, J. P., Rigouard, M., and Seguin, F.,** Produits de radiolyse de faible poids moléculaire formés au cours de l'irradiation gamma de l'amidon de mais: état d'avancement des travaux, in *The Identification of Irradiated Foodstuffs,* Commission of the European Communities, Directorate-General Scientific and Technical Information and Information Management, Luxembourg, 1974, 155.
19. **Delincee, H.,** Radiation Chemistry of Food Components, L47, lecture handouts at IFFIT Training Course on Food Irradiation, International Facility for Food Irradiation Technology, Wageningen.
20. **Tsuji, K.,** Low-dose cobalt-60 irradiation for reduction of microbial contamination in raw materials for animal health products, *Food Technol.*, p.48, February 1983.
21. **Diehl, J. F.,** Thiamin in bestrahlten Lebensmitteln. I. Einfluss verschiedener Bestrahlungsbedingungen und des Zeitablaufs nach der Bestrahlung, *Z. Lebensm. Unters. Forsch.*, 157, 317, 1975.
22. **Diehl, J. F.,** Thiamin in bestrahlten Lebensmitteln. II. Kombinierter Einfluss von Bestrahlung, Lagerung und Erhitzen auf den Thiamingehalt, *Z. Lebensm. Unters. Forsch.*, 158, 83, 1975.
23. **Galetto, W., Kahan, J., Eiss, M., Welbourn, J., Bednarczyk, A. K., and Silberstein, D.,** Irradiation treatment of onion powder: effects on chemical constituents, *J. Food Sci.*, 44, 591, 1979.
24. **Daghir, N. J., Sell, L. J., and Mateos, G. G.,** Effect of gamma irradiation on nutritional value of lentils *(Lens culinaris)* for chicks, *Nutr. Rep. Int.*, 27, 1087, 1983.
25. **Sreenivasan, A.,** Compositional and quality changes in some irradiated food in *Improvement of Food Quality by Irradiation,* International Atomic Energy Agency, Vienna, 1974, 129.

26. **Brynjolfsson, A.,** Food irradiation in the United States, Paper No. E-1, presented at the Int. Meat Res. Congr., Colorado Springs, August 31 to September 5, 1980.
27. Statement on the wholesomeness of irradiated foods by the Surgeon General, Department of the Army, in Radiation Processing of Foods. Hearings before the Subcommittee on Research and Development and Radiation of the Joint Commission on Atomic Energy, Congress of the United States, Washington, D.C., June 9 and 10, 1965, 105.
28. **Phillips, B. J., and Elias, P. S.,** A new approach to investigating the genetic toxicity of processed foods, *Food Cosmet. Toxicol.*, 16, 509, 1979.
29. **Saint-Lebe, L., Berger, G., Muchielli, A., and Coquet, B.,** Evaluation toxicologique de l'amidon de mais irradie, Bilan des travaux en cours, in *Radiation Preservation of Food,* International Atomic Energy Agency, Vienna, 1973, 727.
30. **Truhaut, R., Coquet, B., Guyot, B., Ronaud, M., and Saint-Lebe, L.,** Evaluation toxicologique d'amidon de mais irradié par expérimentation a long terme chez le rat, *Eur. J. Toxicol.*, 9, 347, 1976.
31. **Truhaut, R. and Saint-Lebe, L.,** Différentes voies d'approche pour l'evaluation toxicologique de l'amidon irradié, in *Food Preservation by Irradiation,* Vol. II, International Atomic Energy Agency, Vienna, 1978, 31.
32. **Barna, J.,** Toxicity Tests of Irradiated and Heat Treated Paprika (in Hungarian), Res. Rep., Central Food Research Institute, Budapest, 1973.
33. **Farkas, J.,** Progress in food irradiation — Hungary, *Food Irradiat. Inf.*, 4, 11, 1975.
34. **Barna, J.,** Wholesomeness of an irradiated spice mixture, *Food Irradiat. Inf.*, Suppl. 4, 48, 1975.
35. **Barna, J.,** Preliminary Studies Relating to Investigation of the Wholesomeness of Irradiated Spices, final report to the International Project in the Field of Food Irradiation, Central Food Research Institute, Budapest, 1976.
36. Spices — a new approach, *Food Irradiat. Inf.*, 7, 67, 1977.
37. Teratogenic Study of Irradiated Paprika, Black Pepper and a Spice Mixture, report to the International Project in the Field of Food Irradiation under a subcontract with the Central Food Research Institute, Budapest and the Immunological Department, Biological Institute of the Loránd Eötvös University of Sciences, Alsógöd, Hungary, 1978, IFIP Rep. No. 52, International Project in the Field of Food Irradiation, Karlsruhe, 1979.
38. Central Food Research Institute, Mutagenicity Testing of Irradiated Ground Paprika, interim report IFIP-R44, International Project in the Field of Food Irradiation, Karlsruhe, 1977.
39. **Farkas, J., Andrássy, É., and Incze, K.,** Evaluation of possible mutagenicity of irradiated spices, *Acta Aliment.*, 10, 129, 1981.
40. **Farkas, J., Andrássy, É., and Incze, K.,** Evaluation of possible mutagenicity of irradiated spices, in *5th Int. Congr. of Food Sci. and Technol. — Abstr.*, Kyoto, 1978, 91.
41. **Farkas, J. and Andrássy, É.,** Prophage λ induction (Inductest) of blood of rats fed irradiated spices. *Acta Aliment.*, 10, 137, 1981.
42. **Münzner, R. and Renner, H. W.,** Mutagenicity testing of irradiated onion powder, *J. Food Sci.*, 46, 1269, 1981.
43. **Mittler, S. M.,** Mutagenicity Evaluation of Irradiated Onion Powder, Final Report to McCormick and Co., Inc., Department of Biological Sciences, Northern Illinois University, DeKalb, February 27, 1981.
44. **Mittler, S. and Eiss, M. I.,** Failure of irradiated onion powder to induce sex-linked recessive lethal mutations in *Drosphila melanogaster, Mutat. Res.*, 104, 113, 1982.
45. **Farkas, J. and Andrássy, É.,** A study of possible mutagenicity of irradiated onion powder by *Salmonella* mammalian-microsome tests, *Acta Aliment.*, 10, 209, 1981.
46. **Farkas, J. and Andrássy, É.,** Investigations on the possible genotoxic effects of irradiated onion powder by means of prophage induction (Inductest), *Acta Aliment.*, 11, 245, 1982.
47. Étude des Effects Mutagenes de la Poudre d'Ail apres Irradiation sur des Bacteries *Salmonella typhimurium,* Contrat No. M 106, Institut National des Radioelements et le Laboratoire de Chimie Medicale, Universite de Liege, May 1980.
48. **Ley, F. J.,** Radiation processing of laboratory animal diet, *Radiat. Phys. Chem.*, 14, 677, 1979.
49. **Udes, D., Hiller, H. H., and Juhr, N. C.,** Veränderungen der Rohproteinquantität und -qualität einer Ratten- und Mäusediet durch verschiedene Sterilizations-verfahren, *Z. Versuchstierk.*, 13, 160, 1971.
50. **Barna, J.,** Hygienic investigation on consumptibility of irradiated complete rat diet (in Hungarian), *Izotoptechnika,* 15, 217, 1972.
51. **Ley, F. S., Bledy, J., Coates, M. E., and Paterson, J. S.,** Sterilization of laboratory animal diets using gamma radiation, *Lab. Anim.*, 3, 221, 1969.
52. **Diehl, J. F.,** Vitamin A in bestrahlten Lebensmitteln, *Z. Lebensm. Unters. Forsch.*, 168, 29, 1979.
53. **Diehl, J. F.,** *Lebensm. Wiss. Technol.*, 5, 51, 1972.

54. The Effect of Irradiation upon Spices. II. The Effect of Radiation Treatment on the Lipid Composition of Several Spices, 1st interim report by Central Food Research Institute, Budapest and the Department of Agricultural Chemical Technology, Technical University of Budapest, Hungary, IFIP-R47, International Project in the Field of Food Irradiation, Karlsruhe, April 1978.
55. **Varsányi, I., Liptay, G. Farkas, J., and Petrik-Brandt, E.,** Thermal analysis of spices decontaminated by irradiation, *Acta Aliment.*, 8, 397, 1979.
56. **Reichelt, D. V., Renner, H. W., and Diehl, J. F.,** Long-term animal feeding study for testing the wholesomeness of an irradiated diet with a high content of free radicals (in German) *Ber. Bundesforsch., Lebensm.*, No. 3, 1972.
57. **Renner, H. W.,** Long-term feeding study for testing the wholesomeness of an irradiated diet with a high content of free radicals — 2nd report (in German), *Ber. Bundesforsch. Lebensm.*, No. 1, 1974.
58. **Chaubey, R. C., Kavi, B. R., Barna, J., Chauhan, P. S., and Sundaram, K.,** Cytogenic studies with irradiated ground paprika as evaluated by the micronucleus test in mice, *Acta Aliment.*, 8, 197, 1979.
59. **Diehl, J. F.,** Verminderung von strahleninduzierten Vitamin E- und - B_1 -Verlusten durch Bestrahlung von Lebensmitteln bei tiefen Temperaturen und durch Ausschluss von Luftsauerstoff, *Z. Lebensm. Unters. Forsch.*, 169, 276, 1979.

Chapter 6

LEGISLATIVE ASPECTS AND WORLDWIDE STATUS OF PETITIONS AND CLEARANCE ON IRRADIATED DRY INGREDIENTS

I. INTERNATIONAL STANDARDIZATION OF IRRADIATED FOOD

In order to aid harmonization of national legislation and to facilitate international trade of irradiated food, the Food and Agriculture Organization/World Health Organization (FAO/WHO) Codex Alimentarius Commission has already adopted an International General Standard for Irradiated Foods (see Appendix). The findings and conclusions of the 1980 Joint Expert Committee on the Wholesomeness of Irradiated Foods (JECFI) have been incorporated in a revision of the General Standard, which is now distributed to all 125 member states of Codex Alimentarius for acceptance.

The FAO/WHO Codex Committee on Processed Meat and Poultry Products is evaluating irradiation as an alternative treatment for spices to be used in meat products.

II. PETITIONS AND CLEARANCES ON IRRADIATED PRODUCTS

The national legislative recognition of the safety of irradiated foods follows gradually the achievements at the UN level. By 1986 the number of countries which had granted some sort of clearances for one or more food items or groups of products processed by ionizing energy for human comsumption had risen to 33. Chile and Bangladesh adopted the 1980 JECFI Report and the Codex Alimentarius Recommended International General Standard on Irradiated Food, respectively, and cleared all products listed in those documents. In the German Democratic Republic (G.D.R.) and Yugoslavia, recent regulations gave a practically general clearance to the food irradiation process under specific licensing conditions.[28,29] Governmental committees in Australia and the U.K. are currently considering the adoption of the Codex General Standard by their respective authorities. The Advisory Committee on Irradiated and Novel Foods, set up in 1982 to advise the U.K. Health and Agriculture ministers, has just published its report which endorses the conclusions reached in 1981 by the JECFI report. Reportedly, the U.K. Committee advises that "ionising irradiation up to an overall dose of 10 kGy, correctly applied, provides an efficacious food preservation treatment which will not lead to a significant change in the natural radioactivity of the food or prejudice the safety and wholesomeness of the food....there is no justification on public health grounds for the process not to be permitted within the dose range advised, and the benefits offered present strong grounds for a general clearance to be granted to this extent."[42]

As regards irradiated dry ingredients, dried fruits and vegetables, etc. various clearances have been issued already in Bangladesh, Belgium, Brazil, Bulgaria, Canada, Chile, China, Denmark, France, G.D.R., Hungary, Israel, the Netherlands, New Zealand, Norway, South Africa, the U.S., U.S.S.R., and Yugoslavia (Table 33).

In the U.S. the Food and Drug Administration (FDA) has just published a "final rule" amending its regulations to permit additional uses of ionizing radiation for the treatment of food.[43] The amended Part 179 of the Federal Food, Drug, and Cosmetic Act

1. Permits manufacturers to use irradiation for control of *Trichinella spiralis* in pork carcasses or fresh, nonheat-processed cuts of pork carcasses (minimum dose 0.3 kGy, maximum dose not to exceed 1 kGy)
2. Permits manufacturers to use irradiation at doses not to exceed 1 kGy to inhibit the growth and maturation of fresh foods and to disinfect food to arthropod pests

Table 33
WORLDWIDE STATUS OF CLEARANCES GRANTED ON IRRADIATION OF DRY FOOD INGREDIENTS (AS OF JUNE 1986)

					Radiation source				
							Electron x-rays		
Country	Product	Dose (kGy)	Purpose	Category	^{60}Co	^{137}Cs	≤10 MeV	≤5 MeV	Date of clearance
Bangladesh	Rice and ground rice products	Up to 1	Disinfestation	Unconditional	+	+	+	+	28 Dec. 1983
	Wheat and ground wheat products	Up to 1	Do.	Do.	+	+	+	+	28 Dec. 1983
	Spices	Up to 10	Microbial disinfestation	Do.	+	+	+	+	28 Dec. 1983
Belgium	Paprika	Up to 10	Microbial disinfection	Provisional	+				10 Nov. 1980
	Black/white pepper	Up to 10	Do.	Do.	+				10 Nov. 1980
	Gum arabic	Up to 9	Do.	Do.	+				29 Sept. 1983
	Spices and aromatic substances (78 commodities including onion powder and flakes)	Up to 10	Do.	Do.	+				29 Sept. 1983
	Dried vegetables (7 commodities)	Up to 10	Do.	Do.	+				29 Sept. 1983
Brazil	Wheat flour	Up to 1	Disinfestation	Unconditional	+	+			8 March 1985
	Spices (13 commodities)	Up to 10	Microbial disinfection	Do.	+	+			8 March 1985
Bulgaria	Dry food concentrates	1	Disinfestation	Experimental batches	+				30 April 1972
	Dried fruits	1	Do.	Do.					30 April 1972
Canada	Wheat, flour, whole wheat flour	Up to 0.75	Disinfestation	Unconditional	+				25 Feb. 1969
	Onion powder	Up to 10	Microbial disinfection	Do.	+				10 Dec. 1982

	Spices and dry vegetable seasoning	Up to 10	Do.	Do.	+				Sept. 1984
Chile	Cocoa beans	Up to 5	Microbial disinfection	Unconditional	+	+	+	+	29 Dec. 1982
	Dried dates	Up to 1	Disinfestation	Do.	+	+	+	+	29 Dec. 1982
	Pulses	Up to 1	Do.	Do.	+	+	+	+	29 Dec. 1982
	Wheat and ground wheat products	Up to 1	Do.	Do.	+	+	+	+	29 Dec. 1982
	Rice and ground rice products	Up to 1	Do.	Do.	+	+	+	+	29 Dec. 1982
	Spices and condiments	Up to 10	Microbial disinfection	Do.	+	+	+	+	29 Dec. 1982
China	Peanuts	Up to 0.4	Disinfestation	Do.					30 Nov. 1984
Denmark	Enzyme preparations	Up to 10	Microbiol disinfection	Do.					1983
	Spices	Up to 10	Do.						Dec. 1985
France	Spices and aromatic substances (73 commodities including powdered onion and garlic)	Up to 11	Do.	Do.	+	+	+		10 Feb. 1983 28 Jan. 1986
	Muesli-like cereal product	Up to 10	Do.	Do.	+		+		16 June 1985
	Dried vegetables	Up to 10	Do.	Do.	+		+		16 June 1985
	Gum arabic	Up to 9	Do.	Do.	+		+		16 June 1985
GDR	Spices and condiments	Up to 10	Microbial disinfection	Provisional	+				29 Dec. 1982
Hungary	Mixed spices (black pepper, cumin, paprika, dried garlic for use in sausages)	5	Do.	Experimental batches	+				2 Apr. 1974
	Mixed ingredients for canned hashed meat (wheat flour Na-caseinate, onion and garlic powders, paprika)	5	Do.	Do.	+				20 Nov. 1976
	Spices for sausage production	5	Do.	Test marketing	+				4 Jan. 1982
	Spices for sausage production	5	Do.	Do.	+				28 June 1982
	Spices and condiments including onion and garlic powders	5	Do.	Do.	+				1983

Table 33 (continued)
WORLDWIDE STATUS OF CLEARANCES GRANTED ON IRRADIATION OF DRY FOOD INGREDIENTS (AS OF JUNE 1986)

Country	Product	Dose (kGy)	Purpose	Category	Radiation source				Date of clearance
							Electron x-rays		
					^{60}Co	^{137}Cs	≤10 MeV	≤5 MeV	
	Wheat bran	0.4	Disinfestation	Do.	+				17 July 1984
	Wheat germ, dehydrated	0.4	Do.	Do.	+				11 Oct. 1984
	Spices and condiments	6	Microbial disinfection	Provisional	+				1985
Israel	Spices and condiments (36 commodities including onion powder)	Up to 10	Do.	Unconditional	+		+		6 Mar. 1985
Netherlands	Spices and condiments	8—10	Do.	Experimental batches	+		4 MeV		12 Sept. 1971
	Spices	Up to 10	Do.	Provisional	+		4 MeV		4 Oct. 1974
	Powdered batter-mix	1.5	Do.	Test marketing	+				4 Oct. 1974
	Spices	Up to 10	Do.	Provisional	+		3 MeV		26 June 1975
	Spices	Up to 10	Do.	Do.	+				4 Apr. 1978
	Rice and ground rice products	1	Disinfestation	Do.	+				13 Mar. 1979
	Spices (including onion and garlic powders)	Up to 10	Microbial disinfection	Do.	+				15 Mar. 1979
	Malt	Up to 10	Do.	Do.	+				8 Feb. 1983
	Egg powder	Up to 6	Do.	Do.	+				25 Aug. 1983
	Dry blood proteins	Up to 7	Do.	Do.	+				25 Aug. 1983
	Dried vegetables	Up to 10	Do.	Unconditional	+				27 Oct. 1983
New Zealand	Herbs and spices (one batch)	Up to 10	Do.	Provisional	+				Mar. 1985
Norway	Spices	Up to 10	Do.	Unconditional	+				1982
South Africa	Dried bananas	Up to 5	Disinfestation	Provisional	+				19 Jan. 1977
	Spices		Microbial disinfection	Test marketing	+				1981

	Almonds		Disinfestation	Unconditional	+				Sept. 1984
	Cacao powder	Up to 10	Microbial disinfection	Do.	+				Sept. 1984
	Cheese powder	Up to 10	Microbial disinfection	Do.	+				Sept. 1984
	Dried coconut		Microbial disinfection	Do.	+				Sept. 1984
	Dried vegetables (9 commodities)		Do.	Do.	+				Sept. 1984
	Dried yeasts		Microbial disinfection	Do.	+				Sept. 1984
	Spices and condiments (25 commodities including dried onion and garlic)		Microbial disinfection, disinfestation	Do.	+				Sept. 1984
	Whey powder	Up to 10	Microbial disinfection	Do.	+				Sept. 1984
	Herbal tea (rooibos)		Do.	Do.	+				
U.S.	Wheat and wheat flour	0.2—0.5	Disinfestation	Do.	+	+	5 MeV		1963, 1964, 1966
	Spices and dry vegetable seasonings (38 commodities)	Up to 10	Microbial disinfection, disinfestation	Do.	+	+			5 July 1983 19 June 1984
	Enzyme preparations	Up to 10	Microbial disinfection	Do.	+	+	+	+	10 June 1985
	Dry or dehydrated aromatic vegetable substances	Up to 30	Microbial disinfection	Do.	+	+	+	+	18 Apr. 1986
	Any food	Up to 1	Disinfestation	Do.	+	+	+	+	18 Apr. 1986
U.S.S.R.	Dried fruits	1	Disinfestation	Do.	+				15 Feb. 1966
	Dry food concentrates (buckwheat mush, ground rice, pudding)	0.7	Disinfestation	Do.	+				6 June 1966
Yugoslavia	Dried fruits	Up to 10	Disinfestation	Do.	+				5 Jan. 1985
	Dried vegetables (dried mushrooms)	Up to 10	Microbial disinfection	Do.	+				5 Jan. 1985
	Egg powder	Up to 10	Microbial disinfection	Do.	+				5 Jan. 1985
	Herbal teas	Up to 10	Microbial disinfection	Do.	+				5 Jan. 1985

Table 33 (continued)
WORLDWIDE STATUS OF CLEARANCES GRANTED ON IRRADIATION OF DRY FOOD INGREDIENTS (AS OF JUNE 1986)

					Radiation source				
							Electron x-rays		
Country	Product	Dose (kGy)	Purpose	Category	^{60}Co	^{137}Cs	≤10 MeV	≤5 MeV	Date of clearance
	Onion powder and flakes	Up to 10	Microbial disinfection	Do.	+				5 Jan. 1985
	Pulses (legumes)	Up to 10	Disinfestation	Do.	+				5 Jan. 1985
	Spices and condiments	Up to 10	Microbial disinfection	Do.	+				5 Jan. 1985
	Wheat and ground wheat products	Up to 10	Disinfestation	Do.	+				5 Jan. 1985

Note: The term ''provisional'' indicates that the approval is limited in time and/or quantity.

3. Permits manufacturers to use irradiation at doses not to exceed 10 KGy for microbial disinfection of dry or dehydrated enzyme preparations (including immobilized enzymes)
4. Permits the use of irradiation at doses not to exceed 30 kGy for microbial disinfection of the following dry and dehydrated aromatic vegetable substances: culinary herbs, seeds, spices, teas, vegetable seasonings, and blends of these aromatic vegetable substances; turmeric and paprika may also be irradiated when they are to be used as color additives; the blends may contain sodium chloride and minor amounts of dry food ingredients ordinarily used in such blends

The document published in the *Federal Register* responds to comments on the February 14, 1984 proposed rule.[1]

The growing interest in radiation disinfection is shown by the fact that further petitions for clearances on irradiated spices and herbs are pending or prepared recently at least in Australia, Federal Republic of Germany (F.R.G.), Spain, Sweden, and the U.K. Petitions are also pending on papain, dried blood serum, gelatin, and teas in Belgium; on radiation disinfestation of dried fruits and legumes, radiation of powdered egg white and dehydrated blood plasma in France; on dried vegetables (onions, leeks, garlic, horseradish, celeryroot, carrots, and mushroom) in F.R.G.; and on ginseng in the Korean Republic with a maximum dose of 10 kGy. One company in the U.S. has submitted a petition for use of irradiated fish protein concentrate as a component of pet foods.

III. CONSIDERATIONS ON PACKAGING MATERIALS

Proper packaging is a part of the considerations preceding the approval of the application of radiation treatment for any food product.[32] In selecting appropriate packaging, it is required that packaging materials must (1) be resistant to radiation with respect to their functional protective characteristics, (2) not transmit toxic substances into the food, and (3) not transmit disagreable odors or flavors to the food.

Using rigid containers one has to keep in mind the low penetrating capacity of accelerated electrons and the fact that normal glass undergoes brownish discoloration on radiation. The intensity of this discoloration increases with increasing doses.

Experiments with numerous flexible packaging materials and the wide experience gathered during the decades of sterilization of medical products allows the conclusion that most flexible packaging materials suitable for the use in the food industry are suitable also for irradiated food in the dose range foreseen for dry ingredients.[2-4] Generally, equal or lower amounts of components migrate from the irradiated foils than from the nonirradiated ones.[5]

The U.S. FDA specifically approved a number of flexible packaging materials as food contactants for radiation processing. These approvals are shown in Table 34.

IV. LABELING OF IRRADIATED FOODS

The labeling of irradiated food is a controversial issue. While most countries which issued some kind of regulations on food irradiation require that foods processed by radiation and packaged for retail sale should be prominently marked by a statement such as ''treated by ionizing radiation'' or ''treated with ionizing energy'', opinions are widely differing on labeling ingredients and second generation products produced with irradiated ingredients.[27] No labeling is required for irradiated foods sold in South Africa.[35]

The 1980 FAO/International Atomic Energy Agency (IAEA)/WHO JECFI did not think it necessary ''on scientific grounds to envisage special requirements for quality, wholesomeness and labelling for irradiated foods.''[6] It is worthwhile to note in this context that labeling statements were not required for processes such as methylbromide or ethylene oxide (EtO) treatment.

Table 34
APPROVALS BY THE USFDA FOR RADIATION PRESERVATION OF FOOD PACKAGING MATERIALS (JULY 1971)

Packaging material	Petitioner	Source	Dose (Mrad)	Food additive petition no.	Filling[a]			Regulation[a]		
					Date	Vol.	Page	Date	Vol.	Page
Nitrocellulase coated cellophane	AEC	Gamma	1	1297	2/8/64	29	2318	8/14/64	29	11651
Glassine paper	AEC	Gamma	1	1297	2/8/64	29	2318	8/14/64	29	11651
Wax-coated paperboard	AEC	Gamma	1	1297	2/8/64	29	2318	8/14/64	29	11651
Polypropylene film with or without adjuvants	AEC	Gamma	1	1297	2/8/64	29	2318	8/14/64	29	11651
Ethylene-alkene-1-copolymer	AEC	Gamma	1	1297	2/8/64	29	2318	8/14/64	29	11651
Polyethylene film	AEC	Gamma	1	1297	2/8/64	29	2318	8/14/64	29	11651
Polystyrene film with or without adjuvant substances	AEC	Gamma	1	1297	2/8/64	29	2318	8/14/64	29	11651
Rubber hydrochloride with or without adjuvant substances	AEC	Gamma	1	1297	2/8/64	29	2318	8/14/64	29	11651
Vinylidene chloride-vinyl-chloride copolymer film (Saran wrap)	AEC	Gamma	1	1297	2/8/64	29	2318	8/14/64	29	11651
Polyolefin film with or without adjuvant substances or vinylidene chloride coating (Saran)	AEC	Gamma	1	5M1674	7/30/65	30	9551	3/19/68	33	4659
Polyethylene terephthalate with or without adjuvant substances or vinylidene chloride coatings (Saran) or polyethylene coatings	AEC	Gamma	1	5M1675	7/30/65	30	9551	3/19/68	33	4659
Nylon 11	AEC	Gamma	1	6M1820	9/8/65	30	11400	3/19/68	33	4659
Vinylidene chloride copolymer (Saran) coated cellophane	AEC	Gamma	1	5B1670	2/18/65	30	9116	6/11/65	30	7599
Vegetable parchment	U.S. Army	Gamma	6	5M1622	1/15/65	30	547	3/12/65	30	3354
Polyethylene film with or without adjuvants	U.S. Army	Gamma	6	5M1645	7/21/65	30	9116	6/10/67	32	8360
Polyethylene terephthalate with or without adjuvants	U.S. Army	Gamma	6	5M1645	7/21/65	30	9116	6/10/67	32	8360
Nylon 6 films with or without adjuvants	U.S. Army	Gamma	6	5M1645	7/21/65	30	9116	6/10/67	32	8360
Vinyl chloride-vinyl acetate copolymer films with or without adjuvant	U.S. Army	Gamma	6	5M1645	7/21/65	30	9116	6/10/67	32	8360
Kraft paper	U.S. Army	Gamma	50,000 rad	7M2172	5/23/67	32	7877	7/19/67	32	10567

[a] Date, volume, and page number of *Federal Register* publication.

Compiled by Frank Lebne, U.S. Atomic Energy Commission, Washington, D.C.

More and more national authorities accept the view that specific labeling statements are not required on nonirradiated packaged foods that contain irradiated minor ingredients, i.e., spices.[7] According to the suggestions of the Canadian Bureau of Consumer Affairs, food containing an ingredient treated with ionizing radiation would not be required to indicate in the list of ingredients that the ingredient had been so treated unless the total mass of such ingredients constitutes more than 15% of the total mass of the product.[8] Declaration of radiation treatment on the accompanying bills of lading and on invoices of an irradiated food product is a general requirement as an information for food processors and food trade.

The U.S. FDA currenty requires that foods that are irradiated be labeled to show this fact both at the wholesale and at the retail level.[43] The label and labeling of retail packages of foods irradiated shall bear the following logo

along with either the statement "treated with radiation" or the statement "treated by irradiation" in addition to information required by other regulations. This logo was developed several years ago in the Netherlands, and it has been adopted or its adoption is under consideration in several countries.

The logo shall be placed prominently and conspicuously in conjunction with the required statement. These wording requirements pertaining to the label and labeling of retail packages of food shall expire April 18, 1988, unless extended by the FDA by publication for notice and comment in the *Federal Register.*

For a food, any portion of which is irradiated in conformance with the U.S. regulations, the label and labeling and invoices or bills of lading shall bear either the statement "treated with radiation — do not irradiate again" or the statement "treated by irradiation — do not irradiate again" when shipped to a food manufacturer or processor for further processing, labeling, or packing.

FDA advises that the retail labeling requirement applies only to food that has been irradiated when that food has been sold as such (first generation food), not to food that contains an irradiated ingredient (second generation food) but has not itself been irradiated.

The recommendations of the Codex Alimentarius Commission should also offer possibilities regarding the development of an internationally harmonized approach to regulating the labeling of irradiated foods both for the purposes of consumer information and of providing the food industry and trade as well as the enforcement authorities with appropriate information concerning the conditions and purpose of irradiation. According to the Codex General Standard for Irradiated Foods, labeling of prepackaged foods or of food containing irradiated components is required to be in accordance with the Codex General Standard for the Labeling of Prepackaged Foods (a revised version of this document is under elaboration by the Codex Committee on Food Labeling).[9] The label or accompanying documents of irradiated foods, whether prepacked or bulk, are required to state the fact of irradiation and to give appropriate information as to the identity of the irradiation facility.

The Codex Committee on Food Labeling decided at its March 1985 meeting that irradiated foods should be identified. However, since most countries are currently debating what type of identification is appropriate, the Committee did not stipulate what form the identification should take. The Committee limited its proposal to "A food which has been treated with ionizing radiation energy shall indicate on the label that treatment in close proximity to the name of the food."[34] The Committee also recommended that irradiated ingredients be declared in the list of ingredients or indicated on the label. These recommendations were adopted by the Codex Alimentarius Commission at its July 1985 session, with the stipulation that after 2 years the labeling requirement will be examined to see if it has impeded trade.

FAO and IAEA had jointly established an Advisory Group on regulatory and technological requirements for the authorities of the food irradiation process. This Advisory Group had noted that in response to consumer demands for information or by analogy with other forms of processing governments might require special declarations on the label of irradiated prepackaged foods. The Advisory Group thought, however, that labeling provisions related to irradiated ingredients would be of little value, and irradiated foods present in processed food products should not be so declared on the label unless they were present in amounts which would characterize the product.

V. IDENTIFICATION OF IRRADIATED INGREDIENTS

In general, no reliable routine test is available for simple identification of irradiated foods. In specific cases, however (e.g., when the irradiated ingredient is chemically almost homogeneous, i.e., starch or sugar), specific tests may work.

Measurement of the dose-dependent formation of colored products upon addition of 2-thio-barbituric acid, based on the radiation-induced formation of aldehydes and deoxy-sugars, has been suggested as a method for the identification of irradiated starch.[10-12] Another method for the same purpose may be the reaction of carbonyls with 2,3-dinitro-phenyl-hydrazine (a well-known reaction in carbohydrate chemistry). Conditions of irradiation (temperature and humidity) greatly influence the intensity of coloring obtained with thio-barbituric acid.[40] The optical density (intensity of color reaction) decreases with increasing storage time also as a function of temperature and humidity. Nevertheless, Berger et al.[41] claim that it is possible to detect by this method irradiation of maize starch having been stored for 3 months at 25°C, provided that the irradiation dose had been of at least 0.25 kGy.

According to Scherz,[13] detection of irradiation treatment of flour with doses of 5 kGy and above is possible after concentration of the radiation-induced substances with the aid of liquid-liquid extraction, dialysis, and ion-exchange methods. Deschreider[14] proposed spectropolarimetric and turbidimetric detection of starch depolymerization in flour induced by irradiation doses of 0.25 kGy and above.

Since both production and fate during storage of radiolytic products are influenced by temperature and moisture conditions, these detection methods have rather limited information value.

In chemically multicomponent ingredients, the chemical detection of radiolytic products formed in extremely low concentration makes chemical identification of radiation processing in most of the cases nonfeasible. Some physical tests seem to be more promising in this regard.

Formation and disappearance of free radicals in irradiated spices were investigated by electron-spin-reasonance (ESR) measurements in Norway, Hungary, and the U.S.[15-18] In Hungary, ESR studies were performed on untreated and irradiated samples of paprika powder, ground black pepper, and a spice mixture.[19] Gamma radiation doses from 5 to 50 kGy were applied. In the case of paprika samples, the effect of moisture content on the formation and disappearance of radiation-induced free radicals was also investigated. Shortly after irradiation (on the day of radiation treatment) high amounts of free radicals were detected in irradiated samples, but they diminished upon storage. After a period of 3 weeks, the ESR signals of the irradiated samples approximated those of the controls (Figure 18). Free radicals found in unirradiated ground spices did not disappear during a storage period as long as 1 year. The formation and disappearance of radiation-induced free radicals were found to be strongly affected by the moisture content of samples. If a sample of low moisture content containing a high free radical concentration was placed in an atmosphere of higher moisture content, the free radicals decayed rapidly.

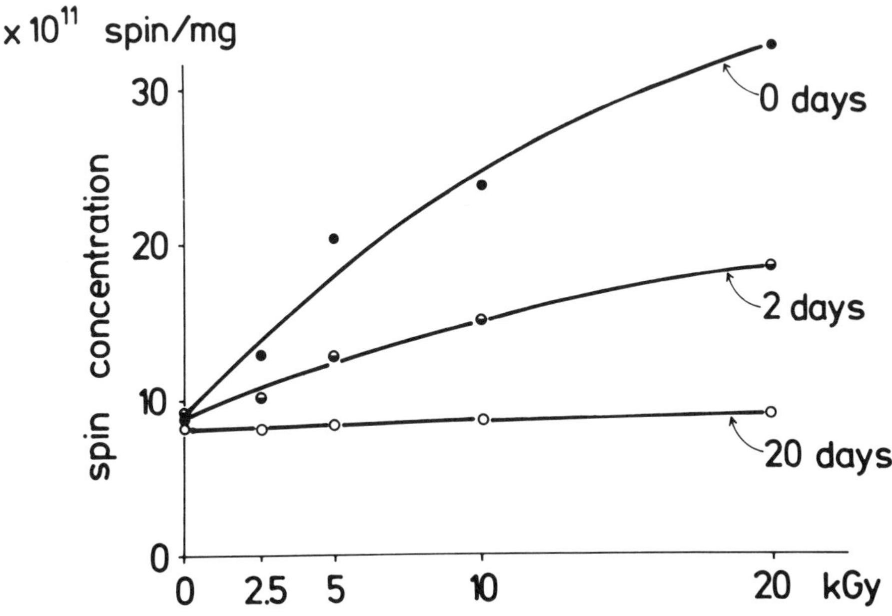

FIGURE 18. The spin concentration in paprika powder as a function of irradiation dose and storage time at room temperature. (From Farkas, J., *Aspects of the Introduction of Food Irradiation in Developing Countries*, International Atomic Energy Agency, Vienna, 1973, 43. With permission.)

In similar ESR studies free radicals produced in dehydrated onions by irradiation decayed within 4 to 5 days at 25°C with a half-life of 0.45 day. Irradiation of sage and black pepper under identical conditions produced free radicals which decayed reaching background level within about 10 days.[18]

Free radical content of cocoa powder was increased considerably by irradiation, and it still could be detected after 6 weeks of postirradiation storage.[20]

Bachman et al.[36] monitored the life time of free radicals in dry gelatin, irradiated at approximately 50 kGy and stored in air at 20°C. Most of these radicals decayed over a few days; some were still detectable after 1 month. Earlier work indicates that they involve rupture of peptide bonds.[37]

One should note that the appearance of free radicals in a foodstuff is not exclusive to radiations, but drying, grinding, etc. are likely to cause the formation of free radicals in detectable quantities, and it is not easy to make a distinction between free radicals formed upon irradiation and those caused by other effects.

Recently new efforts were made in F.R.G. to develop a method for detecting the irradiation of spices. The method reported by Boegl and Heide was also based on the detection of oxidative free radicals with a Luminol solution.[21,22,30] The proposed method is based on the chemiluminescence effect, i.e., on the observation that light is emitted on dissolution or at least suspension of irradiated solids in certain liquids.[23] The light emission can be significantly intensified by the use of a photosensitizer, i.e., the Luminol. The reaction of irradiated (4 or 10 kGy) samples of most spices studied with the luminol reagent adjusted to pH 10 to 11 resulted in an increased chemiluminescence intensity which could be measured instrumentally by a luminescence photometer.[30] The integrated value (unit of measurement: millivolt/second) of the light emission from the start of reaction up to 5 sec after it was measured as well as the maximal light intensity (unit of measurement: millivolt). (Much more light is emitted during the chemiluminescence/lyoluminescence reaction when "pure" and completely soluble solid substances, e.g., salt or sugar, are irradiated. Actually, various dosimetric methods were developed on the basis of lyoluminescence/chemiluminescence of irradiated alkali halides, saccharides, and amino acids.[22,24-26])

Because of uncertainties of the chemiluminescence method, the same group tried recently with rather encouraging results to apply thermoluminescence measurements for detecting treatment of spices with ionizing radiation.[31,38]

Thermoluminescence is the thermally stimulated light emission of solid substances following the removal of excitation, which is caused, for example, by ionizing radiation, i.e., light emission is here a result of heating irradiated solid samples. By measuring the light intensity as a function of temperature, the so-called glow curve is obtained. In the above studies the light emission was induced by a linear heating of spice samples to 300°C within 30 sec in a Harsaw Thermoluminescence Analyzer. After starting the heating cycle, light intensity was recorded and integrated between 60° and 300°C.

The results of these pioneering studies showed that thermoluminescence and chemiluminescence intensities can be very different from one spice to another, and individual spices showed very different dose-response curves. Intensity increases in samples treated with 10 kGy varied between factor 1 (no effect) and approximately 1000 in comparison to untreated samples. In general, the measurement of thermoluminescence was more sensitive than chemiluminescence measurements. It was possible to identify radiation treatment with 10 kGy by at least one of the two methods if irradiation occurred 2 to 3 weeks prior to the examination. In many spices, an identification was possible even 6 to 12 months after irradiation. Comparison of thermoluminescence and chemiluminescence data of γ-irradiated spices is given in Table 35. Time periods in which spice irradiation with 10 kGy was detectable are shown in Table 36.

The authors concluded that in combination with microbiological investigations the luminescence analysis is a simple and rapid identification method for many irradiated dry foodstuffs. While the above studies gave certainly encouraging results, particularly in the absence of other reliable methods for the detection of irradiated food, further studies would be necessary if it were desired that the method should be used as proof for legislatory control purposes, even in the absence of the comparable unirradiated samples and over the expected storage life of the commodities. The multitude of factors influencing the formation of free radicals and their longevity as well as chemiluminescence phenomena induced by UV light and autooxidative processes may reduce the reliability of at least the chemiluminescence signal. Moisture and temperature conditions may have great effects. Albrich et al.[38] reported that after treatment of the irradiated spice samples with saturated water vapor for 18 hr at 25°C, the chemiluminescence intensity was reduced to the value of the unirradiated samples. The thermoluminescence intensity was reduced only about 40% for paprika and 70% for tarragon. Similarly, heat treatments of irradiated samples affected both chemiluminescence and thermoluminescence intensity. After 18-hr treatment at 100°C, irradiated samples could not be identified at all.

According to a recent publication by Mohr and Wichmann,[39] the reduction of viscosity observed in irradiated hydrocolloids (see Chapter 2, Section IV) may be utilized for detection of irradiation of those spices which give normally highly viscous solutions when their suspensions are boiled. The 10% solutions of ground black pepper, mustard seed, green pepper, savory, and caraway seed, respectively, showed marked differences between viscosity of untreated and 8-kGy-irradiated samples.

One should emphasize, however, that even the lack of routine tests for reliable identification of irradiated ingredients should not hamper the application of the process. Effectiveness can be checked by microbiological tests of the processed product and by control of the process itself and record keeping (see Chapter 3, Section IV). This practice is not unique to irradiation. A similar situation exists with heat-sterilized food where bacteriological control is statistically too unreliable and with such heat-pasteurized food where no suitable enzyme

Table 35
COMPARISON OF THERMOLUMINESCENCE (TL) AND CHEMILUMINESCENCE (CL) DATA OF IRRADIATED SPICES[38]

	Luminescence intensity irradiated sample/luminescence intensity unirradiated sample							
	Immediately after irradiation						~4 weeks after irradiation	
	1.5 kGy		4.0 kGy		10 kGy		10 kGy	
Spice	TL	CL	TL	CL	TL	CL	TL	CL
Allspice, whole berry	—	1.5	—	2.3	—	6.4	—	2.5
Allspice, ground	1.3	1.2	2.7	1.5	3.5	2.0	1.0	1.2
Aniseed	—	1.8	—	9.4	—	27.4	—	10.0
Basil	2.4	3.5	4.3	6.6	6.3	6.5	8.5	4.6
Caraway	3.3	1.7	5.9	2.1	7.4	3.0	10.0	3.0
Celery	8.7	15.8	15.5	92.0	18.5	225.0	6.0	23.0
Chili	51.0	1.0	40.1	1.4	76.2	2.6	60.0	2.8
Chive	9.4	1.3	13.3	3.3	29.3	3.4	9.0	2.8
Cinnamon	17.4	184.1	43.8	374.4	50.1	650.0	35.0	370
Cloves, whole flower-bud	—	2.4	—	1.4	—	2.6	—	4.4
Cloves, ground	3.0	1.1	9.0	1.1	10.0	1.3	1.0	1.0
Coriander, single seed	—	6.5	—	11.0	—	16.0	—	15.0
Coriander, ground	6.0	2.1	12.0	4.5	15.0	7.8	3.9	6.3
Cumin	18.8	1.2	26.6	2.3	51.3	7.8	8.0	6.0
Curcuma	330.0	3.1	790.0	5.0	1030	10.4	400.4	5.4
Fennel, single seed	—	7.2	—	19.4	—	34.4	—	17.8
Fennel, ground	2.2	1.3	2.4	1.9	2.5	3.0	1.5	4.0
Garlic	7.5	1.6	19.4	2.4	26.0	5.2	2.8	1.5
Juniper berries, whole berry	—	97.1	—	222.6	—	933.0	—	55.0
Juniper berries, ground	1.4	5.0	1.5	13.5	2.1	31.0	1.6	20.0
Onion	5.4	1.8	7.4	3.4	12.3	9.9	2.0	2.0
Paprika	67.0	—	128.0	—	176.0	2.0	80.5	2.0
Parsley	17.0	1.0	34.3	1.5	44.2	2.1	40.0	2.0
Pepper, black	2.7	1.4	5.0	2.2	15.6	3.8	3.0	2.1
Pepper, white	1.6	1.8	2.3	3.5	5.7	8.1	2.4	2.5
Sage	8.7	1.0	10.7	1.0	19.7	1.0	3.0	1.0
Sesame	—	3.1	—	13.4	—	44.1	—	9.8
Tarragon	29.2	1.7	66.8	4.8	84.1	8.1	31.1	6.0

Note: — = not investigated.

Table 36
TIME PERIODS (DAYS) IN WHICH SPICE IRRADIATION WITH 10 kGy WAS DETECTABLE WITH LUMINESCENCE TECHNIQUES[38]

Spice	Detectability period (days)	
	Chemiluminescence	**Thermoluminescence**
Allspice, whole berry	75	—
Allspice, ground	3	14
Aniseed	≫302	—
Basil	200	>229
Caraway	∼60	∼200
Celery	>313	>200
Chili	77	>195
Chive	50	∼70
Cinnamon	≫95	≫209
Cloves, whole flower-bud	∼200	—
Cloves, ground	0	14
Coriander, single seed	>220	—
Coriander, ground	>126	>80
Cumin	>238	>144
Curcuma	∼200	≫140
Curry	>85	—
Fennel, single seed	>283	—
Fennel, ground	100	∼70
Garlic	3	∼80
Juniper berries, whole berry	≫165	—
Juniper berries, ground	103	21
Onion	37	∼130
Paprika	∼40	∼175
Parsley	8	∼120
Pepper, black	14	∼120
Pepper, white	50	∼50
Sage	0	∼100
Sesame	∼200	—
Tarragon	200	≫190

Note: — = not investigated; positive identification: factor = luminescence-intensity of irradiated sample/luminescence-intensity of unirradiated sample ⩾ 2; ≫ factor higher than 10, > factor lower than 10.

tests have been developed for routine control of process effectiveness. The irradiation process is, for physical reasons, easier to control and check by dosimetry than the traditional processes.

REFERENCES

1. Department of Health and Human Services, Food and Drug Administration, Irradiation in the production, processing and handling of food; proposed rule, 21 CFR part 179. *Fed. Regist.*, 49(31), 5714, 1984.
2. **Mehringer, W.**, Verpackungsprobleme bei der Strahlenkonservierung von Lebensmitteln, *Fette Seifen Anstrichm.*, 71, 516, 1969.
3. **Agarwal, S. R. and Sreenivasan, A.**, Packaging aspects of irradiated flesh foods: present status, a review, *J. Food Technol.*, 8, 27, 1972.
4. **Gopal, N. G. S.**, Effects of radiation sterilization on packaging materials, *Packag. India*, p.1, April/June 1976.
5. **Figge, K. and Freytag, W.**, Einfluss der γ-Bestrahlung von Kunststoffen auf die Auswanderung ihrer Additive in Prüflebensmittel, *Dtsch. Lebensm. Rundsch.*, 73,205,1977.
6. Wholesomeness of Irradiated Food, Report of a Joint FAO/IAEA/WHO Expert Committee, WHO Tech. Rep. Ser. No. 659, World Health Organization, Geneva, 1981.
7. Irradiation labeling, *Food Chem. News*, p. 2, August 26, 1983.
8. Canada proposes treating food irradiation as a process, *Food Chem. News*, p. 13, August 29, 1983.
9. Proposed Draft Revised Text of Recommended International General Standard for the Labelling of Prepackaged Foods, Appendix VII, ALINORM 81/22.
10. **Berger, G., Dauphin, J. F., Athanassiades, H., Saint-Lebe, L., Angel, J. P., Rigonard, M., and Seguin, F.**, Produits de radiolyse de faible poids moléculaire formés au cours de l'irradiation gamma le l'amidon de mais; etat d'advancement des travaux, in *The Identification of Irradiated Foodstuffs*, Commission of the European Communities, Directorate-General Scientific and Technical Information and Information Management, Luxembourg, 1974, 155.
11. **Winchester, R. V.**, Detection of corn starch irradiated with low doses of gamma rays, *Starke*, 25, 230, 1973.
12. **Stewart, A. B. and Winchester, R. V.**, Detection of corn starch irradiated with low doses of gamma rays. III. Positive identification of malonaldehyde in irradiated starch by isolation and characterization as the thiobarbiturate, *Starke*, 27, 9, 1975.
13. **Scherz, H.**, Die Identifizierung von Kohlenhydrathaltigen Lebensmittel durch Analyse der Strahlenspezifischen Substanzen, in *The Identification of Irradiated Foodstuffs*, Commission of the European Communities, Directorate-General Scientific and Technical Information and Information Management, Luxembourg, 1974, 169.
14. **Deschreider, A.R.**, *Lebensm. Wiss. Technol.*, 2, 90, 1969.
15. **Tjaberg, T. B., Underdal, B., and Lunde, G.**, The effect of ionizing radiation on the microbiological content and volatile constituents of spices, *J. Appl. Bacteriol.*, 35, 473, 1972.
16. **Farkas, J.**, Radurization and radicidation of spices, in *Aspects of the Introduction of Food Irradiation in Developing Countries*, International Atomic Energy Agency, Vienna, 1973, 43.
17. **Beczner, J., Farkas, J., Watterich, A., Buda, B., and Kiss, I.**, Study into the identification of irradiated ground paprika, in *The Identification of Irradiated Foodstuffs*, Commission of the European Communities, Directorate-General Scientific and Technical Information and Information Management, Luxembourg, 1974, 255.
18. **Shieh, J. J. and Wierbicki, E.**, Free radicals formation and decay in irradiated spices, IAEA-SM-271/67p, poster presented at the Int. At. Energy Agency/Food and Agric. Organ. Int. Symp. on Food Irradiation Processing, Washington, D.C., March 3 to 8, 1985.
19. The effect of irradiation upon spices, formation and disappearance of radiation-induced free radicals in spices, an ESR study, reported by the Central Food Research Institute, Budapest, Hungary, IFIP-R47, International Project in the Field of Food Irradiation, Karlsruhe, April 1978.
20. **Grünewald, Th. and Münzner, R.**, Strahlenbehandlung von Kakaopulver, *Lebensm, Wiss. Technol.*, 5, 203, 1972.
21. **Boegl, W. and Heide, L.**, Die Messung der Chemilumineszenz von Zimt-, Curry-, Paprika- und Milchpulver als Nachweis einer Behandlung mit ionisierenden Strahlen, ISH-Bericht 32, Bundesgesundheitsamt, Institut für Strahlenhygiene, Neuherberg, November 1983.
22. **Boegl, W. and Heide, L.** Nachweis der Gewürzbestrahlung. Identifizierung gammabestrahlter Gewürze Messung der Chemilumineszenz, *Fleischwirtschaft*, 64, 1120, 1984.

23. **Westmark, T. and Grapengiesser, B.**, Observations of the emission of light dissolution of irradiated solids in certain liquids, *Nature (London),* 188, 395, 1960.
24. **Atari, N. A. and Ettinger, K. V.**, Lyoluminescence of irradiated saccharides, *Radiat. Eff.*, 20, 135, 1973.
25. **Ettinger, K. V. and Puite, K. J.**, Lyoluminescence dosimetry. I. Principles, *Int. J. Appl. Radiat. Isot.*, 33, 1115, 1982.
26. **Puite, K. J. and Ettinger, K. V.**, Lyoluminescence dosimetry. II. State-of-the-art, *Int. J. Appl. Radiat. Isot.* 33, 1139, 1982.
27. **Ladomery, L. G. and Nocera, F.**, Technical and legal aspects relating to the labelling of irradiated foodstuffs, *Food Irradiat. Newsl.* 4, 32, 1980.
28. **Coduro, E.**, Zulassung der Bestrahlung von Lebensmitteln und Bedarfsgegenständen in der Deutschen Demokratischen Republik, *Z. Ges. Lebensm.*, 11. 359, 1984.
29. **Katušin-Ražem, B.**, personal communication.
30. **Heide, L. and Bögl, W.**, Die Messung der Chemilumineszenz von 16 Gewürzen als Nachweis einer Behandlung mit ionisierenden Strahlen, ISH-Heft 53, Bundesgesundheitsamt, Institut für Strahlenhygiene, Neuherberg, September 1984.
31. **Heide, L. and Bögl, W.**, Die Messung der Thermolumineszenz — Ein neues Verfahren zur Identifizierung strahlenbehandelter Gewürze, ISH-Heft 58, Bundesgesundheitsamt, Institut für Strahlenhygiene, Neuherberg, December 1984.
32. **Killoran, J. J.**, Packaging irradiated food, in *Preservation of Food by Ionizing Radiation,* Vol. 2, Josephson, E. S. and Peterson, M. S., Eds., CRC Press, Boca Raton, Fla., 1983, 317.
33. HHS approves rule for food irradiation, *Washington Post,* p. 445, December 13, 1985.
34. **Morrison, R. M. and Roberts, T.**, Food Irradiation: New Perspectives on a Controversial Technology. A Review of Technical, Public Health, and Economic Considerations, Office of Technology Assessment, Congress of the United States, Washington, D.C., December 1985.
35. **De Wet, W. J.**, The South African food irradiation programme: role of government institutions, in *Food Irradiation Processing,* International Atomic Energy Agency, Vienna, 1985, 323.
36. **Bachman, S., Galant, S., Gasyna, Z., Witkowski, S., and Zegota, H.**, Effects of ionizing radiation on gelatin in the solid state, in *Improvement of Food Quality by Irradiation,* International Atomic Energy Agency, Vienna, 1974, 77.
37. **Hayden, G. A., Rodgers, S. L., and Friedberg, F.**, *Arch. Biochem. Biophys.*, 113, 247, 1968.
38. **Albrich, S., Stumpf, E., Heide, L., and Bögl, W.**, *Chemiluminescenz- und Thermolumineszenzmessungen zur Identifizierung strahlenbehandelter Gewürze. Eine Gegenüberstellung beider Verfahren,* ISH-Heft 174, Institut für Strahlenhygiene des Bundesgesundheitsamtes, Neuherberg/München, July 1985.
39. **Mohr, E. and Wichmann, G.**, Vish ositätsernidrigung als Indiz für eine Cobalt bestrahlung von Gewürzen, *Gordian,* 85(5), 96, 1985.
40. **Berger, G. and Saint-Lebe, L.**, Un test d'irradiation de l'amidon de mais, base sur l'emploi de l'acide 2-thiobarbiturique, *Starke,* 21, 205, 1969.
41. **Berger, G., Woodhouse, D. R., and Saint-Lebe, L.**, Un test d'irradiation de l'amidon. Mode operatoire, application a' differents amidons, *Starke,* 24(1), 15, 1972.
42. Go-ahead for irradiation?, *Food Manuf. Int.*, p.5, March/April 1986.
43. Department of Health and Human Services, Food and Drug Administration, Irradiation in the production, processing, and handling of food, final rule, 21 CFR part 179, *Fed. Regist.*, 51(75), 13376, 1986.

Chapter 7

OUTLOOK ON COMMERCIALIZATION

Masefield and Dietz[1] listed the following essential factors to be considered when determining whether a given radiation process is likely to be commercially feasible:

1. Development and funding by industry
2. Distinct technical advantages with radiation
3. Distinct economic advantages with radiation
4. Distinct/unique advantages offered by radiation
5. Consumer need
6. Acceptable to regulatory agencies

Since all categories of this "success" formula are becoming positive for radiation decontamination of dry ingredients in an increasing number of countries, the outlook for commercialization of this part of food irradiation is bright.

With the notable exception of several pure hydrocolloids and ingredients with high lipid content, radiation decontamination of dry products is a feasible process. Its safety is well established, and its economical competitiveness is improving. The procedure is direct, simple, requires no additives, and is highly efficient. Its dose requirement is moderate. In certain cases irradiation may result in considerable savings of energy and labor as compared to alternative decontamination techniques. The microflora surviving radiation decontamination of dry ingredients is sensitized by irradiation to subsequent food processing treatments during utilization of such irradiated products. Although at present it cannot be applied universally, the flavor, texture, or other important technological and sensory properties of the majority of ingredients are not influenced at radiation doses necessary for a satisfactory decontamination, and irradiation does not raise the chemical residue problem either.

It appears that time is on the side of radiation. The demand for dry ingredients of good microbiological quality is ever increasing, and food industries can, therefore, afford the cost of irradiation treatment. Significant progress is made in legislation of the process both nationally and internationally. More clearances on irradiated foods are expected to be granted in the future. With spice irradiation becoming widely acceptable on the international scene, the situation is becoming favorable for commercial exploitation and international trade of these products.

While fewer than ten countries are currently using the food irradiation process commercially and the quantities involved are relatively small, large experimental food/feed irradiation facilities exist in a considerable number of countries, and several multipurpose (semi-) industrial irradiators, which are used partly for processing some food or feed, are also in operation. The growing or renewed interest in the process is shown by the increasing number of demonstration or (semi-) industrial food/feed irradiation facilities that are in the planning, design, or construction stage. A survey of such irradiators is attempted in Table 37. It should be noted that this list is not meant to be exhaustive and some of these irradiation facilities may no longer be in operation and/or in the construction stage.

Irradiation of dry ingredients is now an emerging technology in several industrialized countries. Increasing amounts of these commodities are treated commercially in the recent years, first of all in Holland (Gammaster b.v., Ede, and Pilot Plant for Food Irradiation, Wageningen), in Belgium (IRE-MEDIRIS, Fleurus), and the South African Republic (ISOSTER/Pty/Ltd., Johannesburg).[2,3,7,8] Significant quantities of spices are being irradiated in France by the "CONSERVATOME", in the U.S. (Radiation Technology, Inc., Neutron

Table 37
LARGE EXPERIMENTAL, PILOT-SCALE, OR COMMERCIAL IRRADIATORS IN OPERATION OR PLANNING, DESIGN, OR CONSTRUCTION STAGES, WHICH ARE TO BE USED MAINLY OR OCCASIONALLY FOR PROCESSING FOOD OR FEED (AS OF 1985)

			Estimated source strength (kCi/kW)			
Country	**Location**	**Type of irradiator**	**Nominal**	**Actual (date)**	**Completed in**	**Purpose/ product**
Algeria	CEN-CDTB, Alger	ORIS pilot-scale irradiator, French-made, ^{60}Co (CEA)	30/0.44	9.5/0.14 (1985)	1983	Potatoes & onions
Argentina		Truck-mounted, mobile γ-irradiator, ^{60}Co	15/0.22	5/0.07 (1985)		Demonstration R & D
	Ezeiza, Buenos Aires	Batch-type semi-industrial irradiator, ^{60}Co	1000/14.8	300/4.45 (1985)		Multipurpose, mainly nonfood
	San Luis or Cordoba	Spice irradiator, ^{60}Co	200/2.96		Planning	Irradiation of spices
	Buenos Aires or Puerto Madryn	Fish irradiator, ^{60}Co	250/3.70		Planning	Irradiation of fish
Austria	Seibersdorf	^{60}Co		30/0.44 (1974)		Multipurpose nonfood (lab animal feed)
Australia	Australian AEC, Res. Establishment, Sutherland, N.S.W.	Batch-type irradiator, plaque-source, ^{60}Co		70/1.04 (1985)	1981	Multipurpose
	Ansell Int. Ionizing Energy Div., Sydney	AECL J8900 carrier irradiator, ^{60}Co	750/11.1	300/4.44	1985	Multipurpose (mainly nonfood)
Bangladesh	Inst. Food and Radiat. Biol. (IFRB), Dacca	AECL "Gammabeam-650", ^{60}Co	50/0.74	20/0.30 (1985)	1979	Multipurpose

	Dacca or Chittagong	^{60}Co (Soviet-made)	200/2.96	150/2.23	1987	Multipurpose
Belgium	Mol	RITA, ^{60}Co		59/0.87 (1974)		Multipurpose
	IRE-MEDIRIS, Fleurus	"GAMMIR-II", ^{60}Co	1000/14.8	450/6.67 (1985)	1980/1982	Multipurpose
Brazil	Rio de Janeiro	Portable irradiator, ^{137}Cs	100/0.32	80/0.25 (1985)		Multipurpose
	CENA, Piracicaba	AECL "Gammabeam", ^{60}Co	50/0.74	8/0.12 (1985)		Multipurpose
	Empesa Brazil. de Radiacoes, S.A. (EMBRARAD), São Paulo	AECL JS 7400, ^{60}Co		430/6.37 (1985)		Multipurpose
Bulgaria	Novi Krichim	^{60}Co (Bulgarian design)		10/0.15 (1985)		Multipurpose
	Sofia	^{60}Co		12/0.18 (1985)		Multipurpose
Canada	Quebec	Pilot-scale food irradiator AECL"Gammabeam-651", ^{60}Co	200/2.96	30/0.44 (1985)	1985	Multipurpose
Chile	Santiago	Portable irradiator, ^{137}Cs		100/0.32 (1971)		Multipurpose
	Nucl. Res. Ctr., Lo Aquirre, near Santiago	Shuffle-box, and batch of pallets, Spanish design, ^{60}Co	1000/14.8	50/0.74 (1985)	1977	Multipurpose
China, P.R.	Beijing Radiat. Ctr., Beijing Normal Univ.	LINAC, 5 MeV				Multipurpose
		^{60}Co	1000/14.8	200/2.96 (1985)	1985	Multipurpose
	Shanghai Irradiat. Ctr., Shanghai Inst. Nucl. Res.	^{60}Co	500/7.40	200/2.96 (1985)	1985	Multipurpose (fruits, vegs., dry commod.)
Cuba	Havana	Soviet-made, ^{60}Co	200/2.96		1986	Multipurpose

Table 37 (continued)
LARGE EXPERIMENTAL, PILOT-SCALE, OR COMMERCIAL IRRADIATORS IN OPERATION OR PLANNING, DESIGN, OR CONSTRUCTION STAGES, WHICH ARE TO BE USED MAINLY OR OCCASIONALLY FOR PROCESSING FOOD OR FEED (AS OF 1985)

			Estimated source strength (kCi/kW)			
Country	Location	Type of irradiator	Nominal	Actual (date)	Completed in	Purpose/ product
Denmark	Risö Natl. Lab., Roskilde,	Linear electron accelerator, mod. HRC-712 (Haimson Res. Corp.),	—/10	—/8.8	1975	Multipurpose (mainly nonfood)
	NOVO INDUSTRI	AECL, ^{60}Co		13/0.19 (1985)		Enzyme preps.
Ecuador	Escuela Politecnica Nacional, Quito	^{60}Co, French design	20/0.30			Multipurpose
		LINAC, 6 MeV, Soviet made			1986	Multipurpose
Egypt	Natl. Ctr. Radiat. Res. Technol., Cairo	AECL JS 6500, with addtl. loop for food, ^{60}Co	1000/14.8	270/4.0 (1985)		Multipurpose
		Electron accelerator, 1.5 MeV	—/37.5			Multipurpose (mainly nonfood)
France	"Conservatome" Dagneux, Lyon	"LISA" ^{137}Cs		20/0.06 (1985)		Multipurpose
		"D_2-facility", ^{60}Co	300/4.45	60/0.89 (1985)		Multipurpose (mainly nonfood)
		"D_1-facility", ^{60}Co	1000/14.8	600/8.9 (1985)	1962	Multipurpose (mainly nonfood)
	CEA, Cadarache	Portable irradiator, ^{137}Cs	14/0.04	14/0.04 (1985)	1982	Multipurpose

		^{60}Co	15/0.22	6/0.09 (1985)		Multipurpose
	SODETEG-CARIC, Paris	LINAC, "Circe" 6 MeV	—/7	—/7		Multipurpose (mainly nonfood)
	Arnavaux Market, Marseille	Pilot-scale food irradiator, ^{60}Co	100/1.48	50/0.74 (1986)	1986	Multipurpose
		Pallet irradiator, ^{60}Co	2000/29.6	500-600/ 7.4-8.9 (1986)	1986	Multipurpose
	Société Guyomarc'h, Britanny	"CASSITRON" LINAC, 10 MeV,	—/10		1986	Deboned frozen poultry meat
FRG	Bundesanst. Ernährung, Karlsruhe	Varian V-7703 LINAC 10 MeV	—/6	—/6		Multipurpose
	β-γ-Services (BGS), Wiehl-Bomig	Electron accelator, 3 MeV	2000/29.6		1983	Multipurpose (mainly nonfood)
	Grammaster-Munich CrmsH	AECL pallet irradiator type JS 9000, ^{60}Co				Multipurpose (mainly nonfood)
GDR	Agricultural Cooperative, Weideroda, Leipzig	"Bulk Irradiator" GB Z 81 Central Inst. Isotope Radiat. Res., ^{60}Co		40/0.59 (1985)	1981	Onions
	Spickendorf	Pallet irradiator, ^{60}Co	300/4.45	140/2.07	1986	Multipurpose
	Leipzig	Electron accelerator			Planning	Multipurpose
	VEB PROWIKO, Schönebeck	^{60}Co, batch-type, GBE 82		60/0.89 (1985)	1985	Crude enzyme solutions
Hungary	Inst. Isotopes, Budapest	Type "K-120" panoramic irradiator ^{60}Co	200/2.96	120/1.78 (1985)	1968	Multipurpose
		Bulk irradiator, ^{60}Co Inst. Isotopes	42/0.62		1979	Onions
	AGROSTER Co., Budapest	Pilot food irradiator IPARTERV/Inst. Isotopes, ^{60}Co	300/4.45	100/1.48 (1985)	1971	Multipurpose
		^{60}Co	1000/14.8	500/7.4	Planning	Multipurpose
India	BARC, Trombay	Package irradiator, AECL, ^{60}Co	80/1.18		1967	Multipurpose

Table 37 (continued)
LARGE EXPERIMENTAL, PILOT-SCALE, OR COMMERCIAL IRRADIATORS IN OPERATION OR PLANNING, DESIGN, OR CONSTRUCTION STAGES, WHICH ARE TO BE USED MAINLY OR OCCASIONALLY FOR PROCESSING FOOD OR FEED (AS OF 1985)

Country	Location	Type of irradiator	Estimated source strength (kCi/kW)		Completed in	Purpose/ product
			Nominal	Actual (date)		
		BNL portable irradiator, ^{137}Cs		75/0.24 (1985)		Multipurpose
		Throughflow grain irradiator, AECL, ^{60}Co				Grains
		Onion irradiator, mobile			1985	Onions
Indonesia	PAIR-BATAN, Jakarta	''PANBIT'' BARC, panoramic batch irradiator, ^{60}Co	80/1.18	35/0.52	1978	Multipurpose
		Industrial γ-irradiator, ^{60}Co	400/5.93		1983	Multipurpose (mainly rubber tires)
Iraq	Nucl. Res. Ctr., Tuwaitha, Baghdad	AECL ''Gammabeam-650'', ^{60}Co	50/0.74			Dates, onions
Israel	SOR-VAN Radiation, Ltd., Yavne	AECL type JS 6500, ^{60}Co	1000/14.8	150/2.22 (1985)	1971	Multipurpose
	Soreg. Nucl. Res. Ctr., Yavne	Mobile γ-irradiator (MGI), ^{60}Co		18/0.27 (1985)	1985	Multipurpose
	Matmor Feed-Mill, Asdod	Electron accelator, High Voltage Eng. Corp., U.S., 1.5 MeV	—/75	—/50	1985	Poultry feed
Italy	Casaccia, Rome	^{60}Co		60/0.89 (1974)		Multipurpose
	Fucino	Commercial vegetable irradiator, ^{60}Co	300/4.45		1986	Potatoes, onions, garlic

	Gammatome S.p.a., Como	^{60}Co		150/2.22 (1974)		Multipurpose (mainly nonfood)
	Gammarad, Bologna	^{60}Co		140/2.07 (1974)		Multipurpose (mainly nonfood)
Japan	JAERI, Takasaki	JAERI pilot food irradiator, ^{60}Co		30/0.44 (1985)		Multipurpose
	Agric. Assoc., Shihoro, Hokkaido	Commercial potato irradiator, ^{60}Co	300/4.45	300/4.45 (1974)	1973	Potatoes
Republic of Korea	KAERI, Seoul	^{60}Co	150/2.22		Mid-1970s	Multipurpose
		^{60}Co	500/7.4		Planning	Multipurpose
Malaysia	PUSPATI, Sengalor	^{60}Co	400/5.92		Planning	Multipurpose
		Electron accelator (Japanese-made)			Planning	Multipurpose
Mexico	Ctr. for Nucl. Studies, Mexico City	''Gammabeam-650'' AECL, ^{60}Co	50/0.74	15/0.22		Multipurpose
	Inst. Physics, UNAM, Mexico City	Van de Graff accelerator, 1.5 MeV				Maize disinfestation
The Netherlands	Pilot Plant for Food Irradiation, Wageningen	IFFIT, panoramic batch irradiator plaque source, ^{60}Co	100/1.48	25/0.37 (1985)	1979	Multipurpose
	Pilot Plant for Food Irradiation, Wageningen	Continuous plant, Marsh Ltd. design, ^{60}Co	260/3.70	160/2.37 (1985)	1967	Multipurpose
	GAMMASTER b.v., Ede	AECL pallet irradiator type JS 9000, ^{60}Co	2000/29.6	1200/17.8	1982	Multipurpose
Norway	Inst. for Energy Technology, Kjiller	^{60}Co				Multipurpose (mainly nonfood spices)
Pakistan	Pakistan Inst. of Nucl. Sci. & Technol. (PINSTECH), Islamabad	AECL ''Gammabeam-650'', ^{60}Co	50/0.74			Multipurpose
Peru	IPEN, Lima	^{60}Co (Soviet-made)	200/2.96	100/1.48 (1989)	1989	Multipurpose

Table 37 (continued)
LARGE EXPERIMENTAL, PILOT-SCALE, OR COMMERCIAL IRRADIATORS IN OPERATION OR PLANNING, DESIGN, OR CONSTRUCTION STAGES, WHICH ARE TO BE USED MAINLY OR OCCASIONALLY FOR PROCESSING FOOD OR FEED (AS OF 1985)

Country	Location	Type of irradiator	Estimated source strength (kCi/kW)		Completed in	Purpose/ product
			Nominal	Actual (date)		
Philippines	Philippine AEC, Manila	AECL, "651PT" ^{60}Co		30/0.44 (1985)	1985	Multipurpose
Poland	Technical Univ., Lódz	^{60}Co (Polish design)	25/0.37	20/0.30 (1985)		Multipurpose
South Africa	HEPRO (High Energy Processing) (PTY), Ltd., Tzaneen	AECL type JS 8200, ^{60}Co	400/5.92	100/1.48 (1985)	1982	Fruits, fruit products
	NUCOR, Pelindaba	AECL, package irradiator, JS 6500		320/4.74 (1985)	1984	Multipurpose
	ISO-STER (PTY) Ltd., Kempton Park, Johannesburg	AECL type JS 8900, ^{60}Co	4000/59.2	600/8.89	1981	Multipurpose (dehydr. vegs., spices, herbs)
Spain	Madrid	^{60}Co	500/7.41			Multipurpose
Switzerland	Eidgen. Forschung., Wädenswil	^{60}Co		30/0.44 (1973)		Multipurpose
Taiwan	Taipei	^{60}Co			1981	Multipurpose
Thailand	Office of Atomic Energy for Peace, Bangkok	AECL, "Gammabeam-650" type TR 31, ^{60}Co	50/0.74	30/0.44	1975	Multipurpose
	Bangkok	^{60}Co	500/7.4	100/1.48	Planning	Multipurpose
U.K.	ISOTRON PLC, Swindon	Package irradiator (twin plants), ^{60}Co	3000/44.4	1300/19.3 (1985)	1971	Multipurpose

			1000/14.8	600/8.9 (1985)	1972	(Animal Feed)
	ISOTRON, Reading	^{60}Co		400/5.9 (1974)		Multipurpose
U.S.	Hawaii, Devel. Irradiator, Honolulu	^{60}Co		250/3.7 (1967)	1967	Tropical fruits
	Marine Prod. Devel. Irradiator, Gloucester Technol. Labor	^{60}Co		250/3.7 (1965)		Fishery products
	Savannah, Ga.	Grain product irradiator, ^{60}Co		26.6/0.39 (1971)		Grain products
	Northwestern U.S.	Transportable cesium irradiator /TPCI/ ^{137}Cs	280/0.90	280/0.90	1986	Multipurpose
	California	Agric. commodities irradiator, ^{137}Cs	3000/9.6	3000/9.6	1986	Disinfest. of agric. prod.
	Meat Irradiation Technology Center (MITC)	Pork irradiator, ^{137}Cs	2000-3000/ 6.4-9.6	2000-3000/6.4-9.6		Meats (pork)
	Int. Nutronics Inc., Hawaii	^{60}Co			Planning	Tropical fruits
	Radiat. Technol., West Memphis, Ark.	^{60}Co				Multipurpose (mainly nonfood)
	Radiat. Technol., Burlington, N.C.	Package irradiator, mod. RT-4101, ^{60}Co			1983	Multipurpose
	ISOMEDIX, Parsippany, N.J.	^{60}Co	3000/44.4			Multipurpose
	Int. Nutronics Inc., Irvine, Calif.	^{60}Co				Multipurpose
	Neutron Products, Dickerson	^{60}Co				Multipurpose (mainly nonfood)
	South Jersey Process Technol. Inc., Port of Salem, N.J.	RTI mod. RT-4101-40-48			Planning	Multipurpose
U.S.S.R.	VNIIRT	''Stavrida''RPP-100 ^{137}Cs	91.2/0.29		1969	Fish

Table 37 (continued)
LARGE EXPERIMENTAL, PILOT-SCALE, OR COMMERCIAL IRRADIATORS IN OPERATION OR PLANNING, DESIGN, OR CONSTRUCTION STAGES, WHICH ARE TO BE USED MAINLY OR OCCASIONALLY FOR PROCESSING FOOD OR FEED (AS OF 1985)

Country	Location	Type of irradiator	Estimated source strength (kCi/kW) Nominal	Estimated source strength (kCi/kW) Actual (date)	Completed in	Purpose/ product
	Bogutcharovo, Tula	^{60}Co		136/2.0 (1971)		Multipurpose
	Odessa Port Elevator, RDV, Odessa	Electron accelerator plant; two ELV-2 type accelerators, 1.4 MeV	—/20 kW each		1980	Grain disinfest.
Vietnam	Hanoi	Conveyor type irradiator, ^{60}Co (Soviet-made)		200/2.96 (1985)	1988	Multipurpose
Yugoslavia	"Boris Kidric" Institute, Belgrade	^{60}Co	1000/14.8	220/3.26 (1985)	1978	Multipurpose (mainly nonfood)
	"Ruder Boskovic" Institute, Zagreb	^{60}Co		42/0.62 (1985)		Multipurpose

[a] 67480 Ci of ^{60}Co or 312000 Ci of ^{137}Cs emit 1 kW of γ-radiation.

Products, Inc., and several others), in Hungary (AGROSTER, Budapest), and in Norway. The Novo Industry A/s. in Denmark installed in 1971 a ^{60}Co plant to irradiate powdered enzyme preparations. A regional food irradiation facility (''APIONAL'') is under construction in Marseille, France. Commercial activity of the plant will be started mainly with dry and dehydrated products.[4,9]

With special reference to the role played by many developing countries in producing and marketing spices and herbs, radiation treatment of such commodities is of particular importance to those countries where their export is a valuable contribution to convertible currency earnings. In view of this fact, research and development programs have been launched also in Bangladesh, Chile, Egypt, Ghana, India, Indonesia, Republic of Korea, Malaysia, Nigeria, Peru, Philippines, Sri Lanka, Thailand, Turkey, and Zaire on irradiated spices and herbs.[5]

In view of the requirements of the food industry for ingredients of good microbial quality, and in view of the difficulties in the application of the alternative methods, it is expected that after clearance the relevant industries will transfer step-by-step to the reliable irradiation technology.[1,6]

Worldwide commercialization of the process will greatly depend on general international acceptance of the recommendations of the 1980 Joint Food and Agriculture Organization (FAO)/International Atomic Energy Agency (IAEA)/World Health Organization (WHO) Expert Committee on the Wholesomeness of Irradiated Food and/or the International General Standard for Irradiated Foods and the Recommended International Code of Practice for the Operation of Radiation Facilities for the Treatment of Foods as recommended to national public health authorities by the Codex Alimentarius Commission of the Joint FAO/WHO Food Standards Programme.

REFERENCES

1. **Masefield, J. and Dietz, G. R.,** Food irradiation: the evaluation of commercialization opportunities, *Crit. Rev. Food Sci. Nutri.*, 19, 259, 1983.
2. **Farkas, J.,** Recent developments in the implementation of food irradiation, IFFIT Rep. No. 46a, International Facility for Food Irradiation Technology, Wageningen, August 1984.
3. **Farkas, J.,** Recent Developments on Food Irradiation in Europe and the Middle East, IAEA-SM-271/93, Proc. IAEA/FAO Int. Symp. on Food Irradiat., Washington, D.C., March 4 to 8, 1985, International Atomic Energy Agency, Vienna, 1985, 215.
4. **Henon, Y.,** Past and recent events in food irradiation in France, *Food Irradiat. Newsl.*, 8(1), 41, 1984.
5. **Farkas, J.** Feasibility of Food Irradiation Processes to Developing Countries, IFFIT Rep. No. 35, International Facility for Food Irradiation Technology, Wageningen, November 1982.
6. Spices leading charge into food irradiation, *CRA Info.*, June 1985.
7. **Du Plessis, T. A. and Stevens, R. C. B.,** Marketing of irradiated commodities in South Africa, *Radiat. Phys. Chem.*, 25, 75, 1985.
8. **Van der Linde, H. J. and Brodrick, H. T.,** Commercial experience in introducing radurized foods to the South African market, in *Food Irradiation Processing,* International Atomic Energy Agency, Vienna, 1985, 137.
9. **Laizier, J. and Vuillemey, R.,** Present status and prospects of food irradiation in France, *Food Irradiat. Newsl.*, 9(3), 43, 1985.

GLOSSARY OF TERMS

Absorbed dose: The amount of ionizing radiation (energy) imparted (absorbed) per unit mass of irradiated material. It is measured in gy/rad.

Accelerator: A device for increasing the velocity and energy of charged elementary particles, e.g., electrons or protons, through application of electrical and/or magnetic forces.

Activity: The number of spontaneous nuclear transformations occurring in a given quantity of radioactive nuclide during an incremental interval of time, divided by that interval of time. The SI derived unit is the becquerel (Bq); the traditional special unit is the curie (Ci), use of which is being phased out.

Aerobic microorganisms: Microorganisms that need oxygen for growth. Obligate aerobes cannot survive in the absence of oxygen.

Agar: Dried, purified stems of a seaweed. Chemically, it is a polysaccharide which contains D- and L-galactose residues, some of which are esterified with sulfuric acid; partly soluble and swells in water to form a gel. Used in soups, jellies, ice cream, meat and fish pastes, in bacteriological media, and as a stabilizer for emulsions. Also called agar-agar.

Allspice: Dried fruits of the evergreen *Pimenta officinalis*, also known as pimento.

Anaerobic microorganisms: Microorganisms that grow in the absence of oxygen. Obligate anaerobes cannot survive in the presence of oxygen. Facultative anaerobes normally grow in oxygen, but can also grow in its absence.

Anise: The dried ripe fruit of *Pimpinella anisum* L.

Beam: A stream of particles or electromagnetic radiation, going into a single direction.

Bequerel (Bq): The new unit to measure activity of a radioisotope. 1 Bq is equivalent to 1 disintegration per second. Its relationship to the traditional special unit, the curie (Ci) is: 1 Bq $= 2.7 \times 10^{-11}$ Ci

Biological shield: A mass of absorbing material placed around a radiation source to reduce the radiation to a level that is safe for human beings.

Black pepper: The dried immature berry of *Piper nigrum* L.

Bremsstrahlung: Electromagnetic radiation emitted (as photons) when a fast-moving charged particle (usually an electron) loses energy upon being accelerated and deflected by the electric field surrounding a positively charged atomic nucleus. X-rays produced in X-ray machines are Bremsstrahlung (in German, the term means "braking radiation").

Caraway: The dried ripe fruit of *Carum carvi* L.

Cayenne pepper: The ripe dried fruit of *Capsicum frutescens* L., *C. baccatum* L., or some other small-fruited species of *Capsicum.*

Cardamom: The dried, nearly ripe fruit of *Elettaria cardamomum* Maton.

Celery seed: The dried fruit of *Celeri graveolens* (L.) Britton (*Apium graveolens* L.).

Cinnamon: The dried bark of cultivated varieties of *Cinnamomum zeylanicum* Nees.

Cassia: The dried bark of cultivated varieties of *Cinnamomum cassia* (L.) Blume.

Cloves: The dried flower buds of *Caryophyllus aromaticus* L.

Condiment: Substance used to give relish to food (flavor, distinctive taste, appetizing flavor).

Coriander seed: The dried fruit of *Coriandrum sativum* L.

Cumin seed: The dried fruit of *Cuminum cyminum* L.

Curie (Ci): The unit used formerly to measure activity of a radioisotope. 1 Ci = 3.7×10^{10} disintegrations of atomic nuclei per second. 67.3 kCi of ^{60}Co emit 1 kW energy.

Decoction: An extract prepared by boiling the crude drug in water or by pouring boiling water over it and allowing it to stand.

Digestion: An extract prepared by steeping the crude drug in warm water (35° to 40°C).

Dill seed: The dried fruit of *Anethum graveolens* L.

Directly ionizing particles: Charged particles (electrons, protons, α-particles, etc.) having sufficient kinetic energy to produce ionization by collision.

Disinfestation: Control of the proliferation of insect and other pests.

D_{min}, D_{max}: Minimum and maximum absorbed doses in the product.

Dose distribution: The spatial variation in absorbed dose throughout the product, between the extreme values D_{max} and D_{min}.

Dose meter (Dosimeter): A device, instrument, or system showing a reproducible and measurable response to radiation, used to measure or evaluate the quantity termed absorbed dose.

Dose rate (absorbed dose rate): The increment in absorbed dose during a given time interval. In SI the unit of measurement is the gray per second (Gy/sec); the traditional special unit was the rad per second (rad/sec).

Dose uniformity (dose uniformity ratio): The ratio of maximum to minimum absorbed dose in the product.

Dosimetry: The measurement of radiation quantities, specifically absorbed dose, absorbed dose rate, etc.

Electromagnetic radiation: Radiation consisting of associated and interacting electric and magnetic waves that travel at the speed of light. Examples: light, radiowaves, γ-rays, X-rays.

Electron: A particle possessing a unit negative electrical charge. It has a mass of 1/1840th of that of the hydrogen atoms.

Electron accelerator: A device for imparting large amounts of kinetic energy to electrons.

Electron irradiation: Beams of very highly accelerated free electrons from machines called electron accelerators. The penetration of electrons is limited.

Electronvolt (eV): Energy unit applied to measure the energy of radiations and bond strengths between atoms. One electronvolt is the energy an electron gains when it is accelerated by a potential difference of one volt. Its relation to other energy units is 1 eV = 1.602×10^{-12} ergs, = 1.602×10^{-19} joules, = 3.83×10^{-20} calories. The bond strength between atoms is usually in the range of 0.1 to 10 eV. Light of 500-nm wavelength corresponds to a light quantum with an energy of 2.47 eV. The high energy of ionizing radiations is measured in million electron volts, abbreviated MeV.

Enzyme: A complex protein which acts as catalyst, i.e., brings about or accelerates a specific chemical reaction without itself being used up in the reaction.

Essential oil: A volatile oil obtained from a plant, possessing the smell and other characteristic properties of the plant.

Fennel seed: The dried fruit of cultivated varieties of *Foeniculum vulgare* Hill.

Fenugreek: Seed of *Trigonella foenumgraecum* L.

Free radical: An electrically neutral atom of molecule with an unpaired electron in the outer orbit.

Gamma radiation (γ-rays): High-energy electromagnetic radiation with great penetrating power produced during the spontaneous disintegration of the atomic nucleus of certain radioactive nuclides (radioactive isotopes). The most widely used isotope is ^{60}Co; the wavelength of its γ-rays is approximately 0.001nm. γ-Rays are essentially similar to X-rays, but are usually more energetic and are nuclear in origin.

Garlic: Bulb of *Allium sativum* L.

Ginger: The washed and dried or decorticated and dried rhizome of *Zingiber officinale* Roscoe.

Gray (Gy): The new SI unit of absorbed dose. One gray equals one J/kg, and it corresponds to 100 rad. kGy = 10^3 Gy.

Gum arabic: Hydrocolloid and exudate of some tropical acacias. It is a polysaccharide built from galactose and glucoronic acid units, partly esterified with methanol.

G-value: The yield of chemical changes in an irradiated substance in terms of the number of specified chemical changes produced per 100 eV or per joule of energy absorbed from ionizing radiation.

Half-life, radioactive: The time it takes the activity of a radioisotope to reduce to one half its value, gradually changing into its "daughter element". The half-life of 5.27 years for ^{60}Co means that after 5.27 years only one half of the original ^{60}Co remains, after 10.54 years one half of the half, or one quarter of the ^{60}Co remains intact. This corresponds to a reduction of the activity of a ^{60}Co source by 12.324%/year or 1.096%/month.

Herb: A vascular and nonwoody plant; a plant used for medicinal or culinary purposes.

Indirectly ionizing particles: Unchanged particles (neutrons, photons, etc.) which can liberate directly ionizing particles or can initiate nuclear transformations.

Infusion: An extract obtained by steeping the drug in water.

Ion: An atom or molecule that has lost or gained one or more electrons. By this ionization it becomes electrically charged.

Ionization: The process of adding one or more electrons to or removing one or more electrons from atoms or molecules, thereby creating ions. High temperatures, electrical discharges, or nuclear radiations can cause ionization.

Ionizing radiation: Any radiation consisting of *directly* or indirectly ionizing particles or a mixture of both.

Irradiation: Exposure to radiation.

Irradiator: That part of the radiation facility that houses the radiation source and associated equipment, i.e., the radiation chamber inside the radiation protection shield.

Juniper: Berry of *Juniperus communis* L.

Laurel (bay) leaves: The dried leaves of *Laurus nobilis* L.

Linear accelerator: A long straight tube (or series of tubes) in which charged particles (ordinarily electrons or protons) gain in energy by the action of oscillating electromagnetic fields.

Luminescence: Emission of light at low temperatures.

Mace: The dried arillus of *Myristica fragrans* Houtt.

Maceration: The extraction of constituents from a drug by steeping it in water at room temperature for several hours.

Marjoram: The dried leaves, with or without a small proportion of the flowering tops, of *Majorana hortensis* Moench.

MeV: 1 million (10^6) eV.

Mint: Leaf of *Mentha longifolia* (L.) Hudson.

Mustard flour: The powder from mustard seed, with the hulls largely removed and with or without the removal of a portion of the fixed oil.

Mustard seed: The seed of *Sinapis alba* L. (white mustard), *Brassica nigra* (L.) Koch (black mustard), *G. juncea* (L.) Cosson, or varieties or closely related species of the types of *B. nigra* and *B. juncea.*

Nutmeg: The dried seed of *Myristica fragrans* Houtt, deprived of its tests, with or without a thin coating of lime (CaO).

Onion: Bulb of *Allium cepa* L.

Pallet: A low, portable platform of wood, metal, fiberboard, or combinations thereof used to facilitate handling, storage, and transportation of materials as a unit.

Paprika: The dried ripe fruit of *Capsicum annuum* L.

Pectin: A polysaccharide containing D-galacturonate units. Plant tissues contain protopectins cementing the cell walls together. As fruit ripens, protopectin breaks down to pectin and finally to pectic acid under the influence of certain enzymes. Pectin is the setting agent in jams and jellies. The albedo of oranges, lemons, and apple pomace is a commercial source of pectin. It is used as a gelling agent and as an emulsifier and stabilizer.

Photon: The carrier of a quantum of electromagnetic energy. Photons have an effective momentum, but no mass or electrical charge.

Pimento, allspice: The dried, nearly ripe fruit of *Pimenta officinalis* Lind.

Rad (rad): The special unit of absorbed dose which is being superseded by the gray (Gy). 1 rad $=$ 100 ergs of absorbed energy per gram material, $= 6.24196 \times 10^{13}$ eV/g, $= 10^{-2}$ J/kg, $= 2.389 \times 10^{-6}$ cal/g, $= 10^{-2}$ Gy; krad $= 10^3$ rad, Mrad $= 10^6$ rad.

Radiation dose (absorbed dose): The quantity of the energy of the ionizing radiation absorbed by the irradiated material. With a source of given radiation intensity, the dose applied is directly proportional to the time of exposure. The SI-derived unit of absorbed dose is the gray (Gy); the traditional special unit is the rad, use of which is being phased out.

Radiation facility: The engineering plant housing the radiation source and all the ancillary equipment required for carrying out the radiation process.

Radiation shielding: Reduction of radiation by interposing a shield of absorbing material between any radioactive source and a person, work area, or radiation-sensitive device.

Radiation source: An apparatus or radioactive substance in a suitable support that constitutes the origin of the ionizing radiation (e.g., ^{60}Co source rods in a frame, or an electron accelerator).

Radicidation: Radiation process for destruction of some particular pathogen by ionizing radiation.

Radioactive isotope: An unstable isotope of an element that disintegrates spontaneously, emitting radiation.

Radioactivity: The spontaneous decay or disintegration of an unstable atomic nucleus, usually accompanied by emission of ionizing radiation.

Radiolysis: Chemical decomposition caused by irradiation.

Radurization: Radiation process resulting in extension of storage life by ionizing radiation.

Red pepper: The red, dried, ripe fruit of any species of *Capsicum.*

Relative humidity: The ratio of actual humidity to the maximum humidity which air can retain without precipitation at a given temperature and pressure. Expressed as percent of saturation at a specified temperature.

Rem (roentgen equivalent man): The unit used formerly to measure the dose equivalent of a given exposure to ionizing radiations, taking into account the differing biological effectiveness of different types of radiation. 1 mrem = 10^{-3} rem.

REP (acronym for roentgen equivalent physical): An obsolete unit of absorbed dose of any ionizing radiation with a magnitude of 93 erg/g. It has been superseded by the rad and subsequently by the gray.

Rhizome: Underground stem.

Saffron: The dried stigma of *Crocus sativus* L.

Sage: The dried leaf of *Salvia officinalis* L.

Savory: The dried leaves and flowering tops of *Satureia hortensis* L.

SI: The abbreviation for the International System of Units.

Sievert (Sv): The new SI-unit to measure the dose equivalent of a radiation exposure. 1 Sv corresponds to 100 rem.

Source strength: Of a γ-ray radiation source, the strength defines the activity of the radioactive nuclide source material; it is expressed in becquerels (or curies).

Spices: Aromatic natural vegetable substances or mixtures thereof used for the seasoning of food. The term applies to the product either in the whole form or in the ground form.

Star anise seed: The dried fruit of *Illicium verum* Hook.

Tarragon: The dried leaves and flowering tops of *Artemisia dracunculus* L.

Threshold dose: The minimum dose of radiation that will produce a detectable effect.

Thyme: The dried leaves and flowering tops of *Thymus vulgaris* L.

Tragacanth: White or reddish vegetable gum from certain herbs.

Turmeric: The dried rhizome or bulbous root of *Curcuma longa* L.

Vanilla: Dried pod (fruit) of *Vanilla fragrans* (Salisbury) Ames or *V. tahitensis* Moore.

Volatile oil: An essential oil distilled from plant tissue characterized by the readiness with which it evaporates.

Water activity: The ratio of the water vapor pressure of a food to the vapor pressure of pure water under identical conditions of temperature and pressure. It is a measure of water availability in food for microbial growth.

White pepper: The dried mature berry of *Piper nigrum* L. from which the outer hull (or the outer and inner hulls) has been removed.

X-rays (Roentgen-rays or Bremsstrahlung): High-energy electromagnetic radiation, of the same nature as γ-rays. These rays are called X-rays if they are created outside the nucleus of the atom and γ-rays if they are created inside the nucleus. In X-ray machines high-energy electrons bombard a metal plate, which then emits X-rays.

JOINT FAO/WHO FOOD STANDARDS PROGRAMME
CODEX ALIMENTARIUS COMMISSION

CODEX GENERAL STANDARD FOR IRRADIATED FOODS AND RECOMMENDED INTERNATIONAL CODE OF PRACTICE FOR THE OPERATION OF RADIATION FACILITIES USED FOR THE TREATMENT OF FOODS

INTRODUCTION

The Food and Agriculture Organization (FAO)/World Health Organization (WHO) Codex Alimentarius Commission (the Commission) was established to implement the Joint FAO/WHO Food Standards Programme. Membership of the Commission comprises those Member Nations and Associate Members of FAO and/or WHO which have notified the Organizations of their wish to be considered as Members. By July 1, 1983 122 countries had become Members of the Commission. Other countries which participate in the work of the Commission or of its subsidiary bodies in an observer capacity are expected to become Members in the near future.

The purpose of the Joint FAO/WHO Food Standards Programme is to protect the health of consumers and to ensure fair practices in the food trade; to promote coordination of all food standards work undertaken by international governmental and nongovernmental organizations; to determine priorities and initiate and guide the preparation of draft standards through and with the aid of appropriate organizations; to finalize standards and, after acceptance by governments, publish them in a Codex Alimentarius either as regional or worldwide standards.

At its 15th Session, held in July 1983, the Commission adopted a Codex General Standard for Irradiated Foods and a Recommended International Code of Practice for the Operation of Radiation Facilities used for the Treatment of Foods to be sent to all Member Nations and Associate Members of FAO and/or WHO.

The Codex General Standard for Irradiated Foods was developed in accordance with the Codex Procedure for the Revision and Amendment of Codex Standards by the intergovernmental Codex Committee on Food Additives, which also deals with Food Processing, in close cooperation with the International Atomic Energy Agency (IAEA).

EXPLANATORY NOTES

This Standard takes into account the recommendations and conclusions of the Joint FAO/IAEA/WHO Expert Committees convened to evaluate all available data concerning the various aspects of food irradiation, including the wholesomeness of foods processed by ionizing energy. It also takes into account the recommendations of FAO/IAEA/WHO consultations on legislation and standardization of food irradiation.

This Standard refers only to those aspects which relate to the processing of foods by ionizing energy. It is assumed in this Standard that foods processed by irradiation, like any other foods, will be subject to general food regulations relating to quality, hygiene, weights and measures, and so forth. The provisions of this Standard encompass all foods irradiated up to an overall average dose of 10 kGy or lower. The Standard recognizes that the process of food irradiation has been established as safe for general application to an overall average level of absorbed dose of 10 kGy. The latter value should not be regarded as a toxicological upper limit above which irradiated foods become unsafe; it is simply the level at or below which safety has been established. In setting the overall average dose for the general application of food irradiation, it has been recognized that the required dose to achieve the desired technological effect is governed by ''good irradiation practice''. Applying the appropriate dose level is the key to the technologically and economically proper application of food irradiation.

Despite the many investigations designed to detect physical, chemical, and biological changes in foods subjected to ionizing energy, no satisfactory method for identifying food as having been irradiated has so far been developed. While certain effects can be identified, sufficiently precise methods do not exist for regulatory purposes. Therefore, control of commercial food irradiation can only be performed in the irradiation plant. Consequently, the Standard provides certain mandatory provisions concerning the facilities used and for the control of the process in irradiation plants.

As regards ''labeling'' attention is drawn to the following observation of the 1980 Joint FAO/IAEA/WHO Expert Committee on Wholesomeness of Irradiated Food: ''Irradiated foods would be subject to regulations covering foods generally, and to any specific food standards relating to individual foods. It was, therefore, not thought necessary on scientific grounds to envisage special requirements for the quality, wholesomeness, and labelling of irradiated foods.'' However, there can be a ''technical ground'' for specific applications of food irradiation and the declaration of this fact on the label. For instance, in the case of foods irradiated for the purpose of eliminating pathogens (which should not be stored together with potentially contaminated foods), a statement on the label and/or on the shipping documents of such decontamination treatment would be considered appropriate and informative to manufacturers, traders, and others. The present Standard requires that shipping documents accompanying irradiated foods moving in trade should indicate the fact of irradiation together with relevant information so that good irradiation practice can be verified. The labeling of prepackaged irradiated foods intended for direct sale to the consumer is not covered in this Standard and has to be in accordance with the relevant provisions of the Codex General Standard for the Labelling of Prepackaged Foods. This General Standard is in the process of elaboration.

Members of the Commission are requested to notify the Secretariat of the Codex Alimentarius Commission — Joint FAO/WHO Food Standards Programme of their acceptance of the Codex General Standard for Irradiated Foods, according to paragraph 4 of the General Principles of the Codex Alimentarius (see fifth edition of the Commission's Procedural Manual).

Member Nations and Associate Members of FAO and/or WHO which are not Members of the Commission are also invited to notify the Secretariat if they wish to accept the Codex General Standard for Irradiated Foods.

The Codex General Standard for Irradiated Foods will be published in the Codex Alimentarius as a worldwide Codex Standard when the Commission determines that it is appropriate to do so in the light of acceptances received.

The Recommended International Code of Practice for the Operation of Radiation Facilities used for the Treatment of Foods contained in this publication is intended for the guidance of Governments and is not governed by the acceptance procedure for Codex Standards.

CODEX GENERAL STANDARD FOR IRRADIATED FOODS*

(Worldwide Standard)

1. SCOPE

This standard applies to foods processed by irradiation. It does not apply to foods exposed to doses imparted by measuring instruments used for inspection purposes.

2. GENERAL REQUIREMENTS FOR THE PROCESS

2.1. Radiation Sources

The following types of ionizing radiation may be used:

a. Gamma rays from the radionuclides ^{60}Co or ^{137}Cs

b. X-rays generated from machine sources operated at or below an energy level of 5 MeV

c. Electrons generated from machine sources operated at or below an energy level of 10 MeV

2.2. Absorbed Dose

The overall average dose absorbed by a food subjected to radiation processing should not exceed 10 kGy.**,***

2.3. Facilities and Control of the Process

2.3.1. Radiation treatment of foods shall be carried out in facilities licensed and registered for this purpose by the competent national authority.

2.3.2. The facilities shall be designed to meet the requirements of safety, efficacy, and good hygienic practices of food processing.

2.3.3. The facilities shall be staffed by adequate, trained, and competent personnel.

2.3.4. Control of the process within the facility shall include the keeping of adequate records including quantitative dosimetry.

2.3.5. Premises and records shall be open to inspection by appropriate national authorities.

2.3.6. Control should be carried out in accordance with the Recommended International Code of Practice for the Operation of Radiation Facilities used for the Treatment of Foods (CAC/RCP 19-1979, Rev. 1).

3. HYGIENE OF IRRADIATED FOODS

3.1. The food should comply with the provisions of the Recommended International Code of Practice — General Principles of Food Hygiene (Ref. No. CAC/RCP 1-1969, Rev. 1. 1979) and, where appropriate, with the Recommended International Code of Hygienic Practice of the Codex Alimentarius relative to a particular food.

3.2. Any relevant national public health requirement affecting microbiological safety and nutritional adequacy applicable in the country in which the food is sold should be observed.

4. TECHNOLOGICAL REQUIREMENTS

4.1. Conditions for Irradiation

The irradiation of food is justified only when it fulfills a technological need or where it serves a food hygiene purpose† and should not be used as a substitute for good manufacturing practices.

4.2. Food Quality and Packaging Requirements

The doses applied shall be commensurate with the technological and public health purposes to be achieved and shall be in accordance with good radiation processing practice. Foods to be irradiated and their packaging materials shall be of suitable quality, acceptable hygienic condition, and appropriate for this purpose and shall be handled, before and after irradiation, according to good manufacturing practices taking into account the particular requirements of the technology of the process.

5. REIRRADIATION

5.1. Except for foods with low moisture content (cereals, pulses, dehydrated foods, and other such commodities) irradiated for the purpose of controlling insect reinfestation, foods irradiated in accordance with sections 2 and 4 of this standard shall not be reirradiated.

* Revised version of the Recommended International General Standard for Irradiated Foods (CAC/RS 106-1979).

** For measurement and calculation of overall average dose absorbed see Annex A of the Recommended International Code of Practice for the Operation of Radiation Facilities used for Treatment of Foods (CAC/RCP 19-1979, Rev. 1).

***The wholesomeness of foods, irradiated so as to have absorbed an overall average dose of up to 10 kGy, is not impaired. In this context the term "wholesomeness" refers to safety for consumption of irradiated foods from the toxicological point of view. The irradiation of foods up to an overall average dose of 10 kGy introduces no special nutritional or microbiological problems (Wholesomeness of Irradiated Foods, Report of a Joint FAO/IAEA/WHO Expert Committee, Tech. Rep. Ser. 659, World Health Organization, Geneva, 1981).

† The utility of the irradiation process has been demonstrated for a number of food items listed in Annex B to the Recommended International Code of Practice for the Operation of Radiation Facilities used for the Treatment of Foods.

5.2. For the purpose of this standard food is not considered as having been reirradiated when: (1) the food prepared from materials which have been irradiated at low dose levels, e.g., about 1 kGy, is irradiated for another technological purpose; (2) the food, containing less than 5% of irradiated ingredient, is irradiated; or when (3) the full dose of ionizing radiation required to achieve the desired effect is applied to the food in more than one installment as part of processing for a specific technological purpose.

5.3. The cumulative overall average dose absorbed should not exceed 10 kGy as a result of reirradiation.

6. LABELING

6.1. Inventory Control

For irradiated foods, whether prepackaged or not, the relevant shipping documents shall give appropriate information to identify the registered facility which has irradiated the food, the date(s) of treatment, and lot identification.

6.2. Prepackaged Foods Intended for Direct Consumption

The labeling of prepackaged irradiated foods shall be in accordance with the relevant provisions of the Codex General Standard for the Labelling of Prepackaged Foods.*

6.3. Foods in Bulk Containers

The declaration of the fact or irradiation shall be made clear on the relevant shipping documents.

* Under revision by the Codex Committee on Food Labelling.

RECOMMENDED INTERNATIONAL CODE OF PRACTICE FOR THE OPERATION OF IRRADIATION FACILITIES USED FOR THE TREATMENT OF FOODS*

1. INTRODUCTION

This code refers to the operation of irradiation facilities based on the use of either a radionuclide source (^{60}Co or ^{137}Cs) or X-rays and electrons generated from machine sources. The irradiation facility may be of two designs, either "continuous" or "batch" type. Control of the food irradiation process in all types of facility involves the use of accepted methods of measuring the absorbed radiation dose and of the monitoring of the physical parameters of the process. The operation of these facilities for the irradiation of food must comply with the Codex recommendations on food hygiene.

2. IRRADIATION PLANTS

2.1. Parameters

For all types of facility the doses absorbed by the product depend on the radiation parameter, the dwell time or the transportation speed of the product, and the bulk density of the material to be irradiated. Source product geometry, especially distance of the product from the source and measures to increase the efficiency of radiation utilization, will influence the absorbed dose and the homogeneity of dose distribution.

2.1.1. Radionuclide Sources

Radionuclides used for food irradiation emit photons of characteristic energies. The statement of the source material completely determines the penetration of the emitted radiation. The source activity is measured in Becquerel (Bq) and should be stated by the supplying organization. The actual activity of the source (as well as any return or replenishment of radionuclide material) shall be recorded. The recorded activity should take into account the natural decay rate of the source and should be accompanied by a record of the date of measurement or recalculation. Radionuclide irradiators will usually have a well-separated and shielded depository for the source elements and a treatment area which can be entered when the source is in the safe position. There should be a positive indication of the correct operational and of the correct safe position of the source which should be interlocked with the product movement system.

2.1.2. Machine Sources

A beam of electrons generated by a suitable accelerator, or after being converted to X-rays, can be used. The penetration of the radiation is governed by the energy of the electrons. Average beam power shall be adequately recorded. There should be a positive indication of the correct setting of all machine parameters which should be interlocked with the product movement system. Usually a beam scanner or a scattering device (e.g., the converting target) is incorporated in a machine source to obtain an even distribution of the radiation over the surface of the product. The product movement, the width and speed of the scan, and the beam pulse frequency (if applicable) should be adjusted to ensure a uniform surface dose.

2.2. Dosimetry and Process Control

Prior to the irradiation of any foodstuff certain dosimetry measurements** should be made, which demonstrate that the process will satisfy the regulatory requirements. Various techniques for dosimetry pertinent to radionuclide and machine sources are available for measuring absorbed dose in a quantitative manner.***

Dosimetry commissioning measurements should be made for each new food, irradiation process, and whenever modifications are made to source strength or type and to the source product geometry.

Routine dosimetry should be made during operation and records kept of such measurement. In addition, regular measurements of facility parameters governing the process, i.e., transportation speed, dwell time, source exposure time, and machine beam parameters, can be made during the facility operation. The records of these measurements can be used as supporting evidence that the process satisfies the regulatory requirements.

3. GOOD RADIATION PROCESSING PRACTICE

Facility dosing should attempt to optimalize the dose uniformity ratio, to ensure appropriate dose rates, and, where necessary, to permit temperature control during irradiation (e.g., for the treatment of frozen food) and also control of the atmosphere. It is also often necessary to minimize mechanical damage to the product during transportation irradiation and storage and desirable to ensure the maximum efficiency in the use of the irradiator. Where the food to be irradiated is subject to special standards for hygiene or temperature control, the facility must permit compliance with these standards.

4. PRODUCT AND INVENTORY CONTROL

4.1. The incoming product should be physically separated from the outgoing irradiated products.

* Revised version of the Recommended International Code of Practice for the Operation of Radiation Facilities used for the Treatment of Foods (CAC/RCP 19-1979).

** See Annex A to this Code.

***Detailed in the Manual of Food Irradiation Dosimetry IAEA, Vienna, 1977, Tech. Rep. Ser. No. 178.

4.2. Where appropriate, a visual color change radiation indicator should be affixed to each product pack for ready identification of irradiated and nonirradiated products.

4.3. Records should be kept in the facility record book which show the nature and kind of product being treated, its identifying marks if packed or, if not, the shipping details, its bulk density, the type of source or electron machine, the dosimetry, the dosimeters used and details of their calibration, and the date of treatment.

4.4. All products shall be handled, before and after irradiation, according to accepted good manufacturing practices taking into account the particular requirements of the technology of the process.* Suitable facilities for refrigerated storage may be required.

ANNEX A

Dosimetry

1. The Overall Average Absorbed Dose

It can be assumed for the purpose of the determination of the wholesomeness of food treated with an overall average dose of 10 kGy or less that all radiation chemical effects in that particular dose range are proportional to dose.

The overall average dose, $\bar{D}$ is defined by the following integral over the total volume of the goods — $\bar{D} = 1/M \int \rho\,(x,y,z) \cdot d\,(x,y,z) \cdot dV$ where M = the total mass of the treated sample; ρ = the local density at the point (x, y, z); d = the local absorbed dose at the point (x, y, z); and dV = dx dy dz the infinitesimal volume element which in real cases is represented by the volume fractions.

The overall average absorbed dose can be determined directly for homogeneous products or for bulk goods of homogeneous products or for bulk goods of homogeneous bulk density by distributing an adequate number of dose meters strategically and at random throughout the volume of the goods. From the dose distribution determined in this manner an average can be calculated which is the overall average absorbed dose.

If the shape of the dose distribution curve through the product is well determined the positions of minimum and maximum dose are known. Measurements of the distribution of dose in these two positions in a series of samples of the product can be used to give an estimate of the overall average dose. In some cases the mean value of the average values of the minimum ($\bar{D}_{min}$) and maximum ($\bar{D}_{max}$) dose will be a good estimate of the overall average dose, i.e., in these cases, overall average dose $\approx \bar{D}_{max} + \bar{D}_{min}/2$.

2. Effective and Limiting Dose Values

Some effective treatment, e.g., the elimination of harmful microorganisms or a particular shelflife extension or a disinfestation, requires a minimum absorbed dose. For other applications, too high an absorbed dose may cause undesirable effects or an impairment of the quality of the product.

The design of the facility and the operational parameters have to take into account minimum and maximum dose values required by the process. In some low dose applications it will be possible within the terms of section 3 on Good Radiation Processing Practice to allow a ratio of maximum to minimum dose of greater than 3.

With regards to the maximum dose value under acceptable wholesomeness considerations and because of the statistical distribution of the dose a mass fraction of product of at least 97.5% should receive an absorbed dose of less than 15 kGy when the overall average dose is 10 kGy.

3. Routine Dosimetry

Measurement of the dose in a reference position can be made occasionally throughout the process. The association between the dose in the reference position and the overall average dose must be known. These measurements should be used to ensure the correct operation of the process. A recognized and calibrated system of dosimetry should be used.

A complete record of all dosimetry measurements including calibration must be kept.

4. Process Control

In the case of a continuous radionuclide facility it will be possible to automatically make a record of transportation speed or dwell time together with indications of source and product positioning. These measurements can be used to provide a continuous control of the process in support of routine dosimetry measurements.

In a batch operated radionuclide facility, automatic recording of source exposure time can be made, and a record of product movement and placement can be kept to provide a control of the process in support of routine dosimetry measurements.

In a machine facility a continuous record of beam parameters, e.g., voltage, current, scan speed, scan width, pulse repetition, and a record of transportation speed through the beam, can be used to provide a continuous control of the process in support of routine dosimetry measurements.

* See Annex B to this Code.

ANNEX B

Examples of Technological Conditions for the Irradiation of Some Individual Food Items Specificially Examined by the Joint FAO/IAEA/WHO Expert Committee

This information is taken from the Reports of the Joint FAO/IAEA/WHO Expert Committees on Food Irradiation (WHO Tech. Rep. Ser. 604, 1977 and 659, 1981) and illustrates the utility of the irradiation process. It also describes the technological conditions for achieving the purpose of the irradiation process safely and economically.

1. CHICKEN *(Gallus domesticus)*

1.1. Purposes of the Process

The purposes of irradiating chicken are

To prolong storage life

To reduce the number of certain pathogenic microorganisms, i.e., *Salmonella,* from eviscerated chicken

1.2. Specific Requirements

1.2.1. Average Dose: for 1. and 2., up to 7 kGy.

2. COCOA BEANS (*Theobroma cacao*)

2.1. Purposes of the Process

The purposes of irradiating cocoa beans are

To control insect infestation in storage

To reduce microbial load of fermented beans with or without heat treatment

2.2. Specific Requirements

2.2.1. Average Dose: for 1., up to 1 kGy; for 2., up to 5 kGy.

2.2.2. Prevention of Reinfestation: Cocoa beans, whether prepackaged or handled in bulk, should be stored as far as possible under such conditions as will prevent reinfestation and microbial recontamination and spoilage.

3. DATES *(Phoenix dactylifera)*

3.1. Purpose of the Process

The purpose of irradiating prepackaged dried dates is to control insect infestation during storage.

3.2. Specific Requirements

3.2.1. Average Dose: up to 1 kGy.

3.2.2. Prevention of Reinfestation: Prepackaged dried dates should be stored under such conditions as will prevent reinfestation.

4. MANGOES (*Mangifera indica*)

4.1. Purposes of the Process

The purposes of irradiating mangoes are

To control insect infestation

To improve keeping quality by delaying ripening

To reduce microbial load by combining irradiation and heat treatment

4.2. Specific Requirements

4.2.1. Average Dose: up to 1 kGy.

5. ONIONS (*Allium cepa*)

5.1. Purpose of the Process

The purpose of irradiating onions is to inhibit sprouting during storage.

5.2. Specific Requirement

5.2.1. Average Dose: up to 0.15 kGy.

6. PAPAYA (*Carica papaya* L.)

6.1. Purpose of the Process

The purpose of irradiating papaya is to control insect infestation and to improve its keeping quality by delaying ripening.

6.2. Specific Requirements

6.2.1. Average Dose: up to 1 kGy.

6.2.2. Source of Radiation: The source of radiation should be such as will provide adequate penetration.

7. POTATOES (*Solanum tuberosum* L.)

7.1. Purpose of the Process

The purpose of irradiating potatoes is to inhibit sprouting during storage.

7.2. Specific Requirement

7.2.1. Average Dose: up to 0.15 kGy.

8. PULSES

8.1. Purpose of the Process

The purpose of irradiating pulses is to control insect infestation in storage.

8.2. Specific Requirement

8.2.1. Average Dose: up to 1 kGy.

9. RICE (*Oryza* species)

9.1. Purpose of the Process

The purpose of irradiating rice is to control insect infestation in storage.

9.2. Specific Requirements

9.2.1. Average Dose: up to 1 kGy.

9.2.2. Prevention of Reinfestation: Rice, whether prepackaged or handled in bulk, should be stored as far as possible under such conditions as will prevent reinfestation.

10. SPICES AND CONDIMENTS, DEHYDRATED ONIONS, ONION POWDER

10.1. Purposes of the Process

The purposes of irradiating spices, condiments, dehydrated onions, and onion powder are

To control insect infestation

To reduce microbial load

To reduce the number of pathogenic microorganisms

10.2. Specific Requirement

10.2.1. Average Dose: for 1., up to 1 kGy; for 2. and 3., up to 10 kGy.

11. STRAWBERRY (*Fragaria* species)

11.1. Purpose of the Process

The purpose of irradiating fresh strawberries is to prolong the storage life by partial elimination of spoilage organisms.

11.2. Specific Requirement

11.2.1. Average Dose: up to 3 kGy.

12. TELEOST FISH AND FISH PRODUCTS

12.1. Purposes of the Process

The purposes of irradiating teleost fish and fish products are

To control insect infestation of dried fish during storage and marketing

To reduce microbial load of the packaged or unpackaged fish and fish products

To reduce the number of certain pathogenic microorganisms in packaged or unpackaged fish and fish products

12.2. Specific Requirements

12.2.1. Average Dose: for 1., up to 1 kGy; for 2. and 3., up to 2.2 kGy.

12.2.2. Temperature Requirement: During irradiation and storage the fish and fish products referred to in 2. and 3. should be kept at the temperature of melting ice.

13. WHEAT AND GROUND WHEAT PRODUCTS (*Triticum* species)

13.1. Purpose of the Process

The purpose of irradiating wheat and ground wheat products is to control insect infestation in the stored products.

13.2. Specific Requirements

13.2.1. Average Dose: up to 1 kGy.

13.2.2. Prevention of Reinfestation: These products, whether prepackaged or handled in bulk, should be stored as far as possible under conditions as will prevent reinfestation.

INDEX

F

G

H

N

O

P

T

U

V

W

X